Improving Your Health with Calcium and Phosphorus

by Ruth Adams

and

Frank Murray

Preventive Health
Library
Series

Larchmont Books
New York

NOTICE: This book is meant as an informational guide for the prevention of disease. For conditions of ill-health, we recommend that you see a physician, psychiatrist or other professional licensed to treat disease. These days, many medical practitioners are discovering that a strong nutritional program supports and fortifies whatever therapy they may use, as well as effectively preventing a recurrence of the illness.

First printing: June, 1978

IMPROVING YOUR HEALTH WITH CALCIUM AND PHOSPHORUS

ISBN 0-915962-25-X

Printed in the United States of America

LARCHMONT BOOKS
6 East 43rd Street
New York, N.Y. 10017
212-949-0800

Contents

CHAPTER 1

The Importance of Calcium and Phosphorus

SOME NEW FINDINGS about the great importance of calcium have recently come to light. In many cases, they disprove ideas and theories which have been stock in trade for official nutrition experts for many years. Without exception they seem to show that we need far more calcium than anyone has hitherto recommended and that it is all but impossible to get too much.

According to *Executive Health,* November, 1977, more than one-fourth of all American women over 60 have enough osteoporosis (the soft-bone disease) to produce deformity, pain and loss of height. This deficiency disease causes the body to shrink. Your height decreases.

You might think that osteoporosis is a very visible disorder involving chiefly the back, neck and joints. But, as we detail in another chapter, **it seems that the first bones actually robbed of their calcium during calcium deficiency are the jawbones, in which teeth are embedded.** Periodontal disease, which we also call gingivitis or pyorrhea, is a disorder in which pockets of infection appear around the roots of teeth; gums bleed and soon teeth begin to loosen in their sockets. We are told that far more teeth are lost to this disease than to decay, for loose teeth must be extracted.

Executive Health tells us that one researcher at Cornell University, Dr. Lennart Krook, studying bone diseases of animals, theorized that one of the causes was simply not getting enough calcium. By feeding the animals diets deficient in calcium, he was able to reproduce the bone diseases he saw in his veterinary practice. Testing human beings with severe periodontal (gum) disease, Dr. Krook gave each of them 500 milligrams of calcium twice a day. Inflammation around teeth disappeared, teeth tightened up once again. In one case, a woman whose dentist had been treating her for pyorrhea was amazed when he found her condition greatly improving. Two years later another dentist could not believe she had ever had the disorder. She had been taking large doses of calcium.

In Finland, says *Executive Health*, where average daily intake of calcium is more than 1,300 milligrams and in the Sudan where it may be as high as 2,000 milligrams, both pyorrhea and spinal diseases are rare. In parts of India where dairy products are not used, gum disease starts in early childhood and afflicts almost everyone.

Here are reports on studies done by two New York scientists. **They studied 4,000 aging people and found that they invariably tend to lose bone as they age.** Even at the age of 25, 10 to 15 per cent of healthy "normal" people had lost bone density, showing significant loss of minerals.

Unwise reducing diets are partly responsible, especially in women. During pregnancy and breast-feeding, diets will probably not be rich enough in calcium to provide for both mother and baby, especially if the mother was short on calcium to start out with. So calcium supplements should be taken. In menopause, too, women lose calcium from bones due to the loss of hormones which occurs at this time.

Calcium, the most abundant mineral in the body, comprises 1.5 to 2 per cent of the weight of an adult's body, according to Milicent L. Hathaway and Ruth M.

Leverton in *Food*, the Yearbook of Agriculture, 1959. Calcium is usually associated with phosphorus, which is 0.8 to 1.1 per cent of the body weight. A person who weighs 154 pounds would have 2.3 to 1.7 pounds of phosphorus in his body.

About 99 per cent of the calcium and 80 to 90 per cent of the phosphorus are in the bones and the teeth, *Food* continues. The rest is in the soft tissues and body fluids and is highly important to their normal functioning.

Calcium is essential for the clotting of blood, the action of certain enzymes, and the control of the passage of fluids through the cell walls. The right proportion of calcium in the blood is responsible for the alternate contraction and relaxation of the heart muscle.

The irritability of the nerves is increased when the amount of calcium in the blood is below normal, *Food* says.

Further, calcium in a complex combination with phosphorus gives rigidity and hardness to the bones and teeth.

Phosphorus is an essential part of every living cell. It takes part in the chemical reactions with proteins, fats and carbohydrates to give the body energy and vital materials for growth and repair. It helps the blood neutralize acid and alkali.

Both calcium and phosphorus are essential for the work of the muscles and for the normal response of nerves to stimulation, the yearbook reports.

"The human embryo at 12 weeks contains about 0.2 gram of calcium and 0.1 gram of phosphorus. (There are 28.4 grams in an ounce.) The values are 5.5 and 3.4 grams, respectively, for these two minerals by the 28th week, and 11 and 7 grams by the 34th week. The most rapid increase in the calcium and phosphorus content of the unborn child occurs from the 34th week to the 40th week," *Food* says.

"One-half of the total calcium and more than one-third of the total phosphorus in the baby's body at birth are deposited during the last six weeks. The baby's body contains about 23 grams of calcium and 13 grams of phosphorus at birth." (One gram is equal to 1,000 milligrams).

The calcium content of the body increases faster in relation to size during the first year of life than at any other time. About 60 grams of calcium are added. A child is depositing only about 20 grams a year when he is four or five years old and weighs about 40 pounds. He may be depositing as much as 90 grams a year when he is 13 to 14 years old and weighs 110 pounds. He will deposit more if he weighs more, *Food* says.

Of course, all these gains in calcium content depend on an adequate supply of calcium in the diet and the ability of the body to use it for normal growth.

The percentage as well as the total amount of calcium and phosphorus increases during growth. The infant's body is about 0.8 per cent calcium and the adult's is about twice as much—1.5 to 2 per cent. The phosphorus content of the body increases from 0.4 per cent at birth to 0.8 to 1.1 percent in adulthood, *Food* adds.

The yearbook tells us that bone is composed of tiny, complex crystals of calcium and phosphorus, which are set in honeycomb fashion around a framework of softer protein material, called the organic matrix.

The crystals contain about twice as much calcium as phosphorus. They also contain oxygen and small amounts of hydrogen and other minerals.

"The honeycomb structure gives strength and an enormous surface area to a small amount of bone material—as much as 3,100 square yards to one ounce. Connecting canals containing blood and lymph vessels, nerves and bone marrow pass throughout the matrix and bone crystals. Intercellular fluid surrounds the crystals and keeps them supplied with the materials for repair," *Food* says.

"Crystals of the same kind are deposited to make the enamel and dentin of the teeth. The crystals are larger, however, than those in bone. That may be the reason why enamel and dentin are harder than bone.

"Phosphorus and calcium are of equal importance

in the bones," the USDA yearbook continues. "Phosphorus is involved in ossification or calcification just as much as calcium. When bone is formed, phosphorus is deposited with the calcium. When the bone loses calcium (by decalcification), it also loses phosphorus. They are closely associated in blood and foods. Phosphorus is included, therefore, even though it is not named each time that calcium is mentioned."

We are told that the major change during growth is in the size and the compactness of the bone material. The shape of the bones in a young child is much the same as it will be when he is an adult. The bones in an infant are like firm cartilage and have a low content of calcium and phosphorus. They become firmer as these minerals are deposited in and around the cartilage. This process of bone building is called ossification or calcification. The bones and teeth are said to ossify or to calcify.

Certain bones in the wrist and ankle and the permanent teeth do not begin to calcify until after birth. Groups of specialized cells that are present at birth have the ability to deposit calcium and phosphorus around them and thus become bones and teeth. They are called ossification centers and tooth buds, *Food* states.

Further, eight small bones in the wrist are mere ossification centers at birth. Two of them usually are calcified in the first year, one in the third year, two in the fifth year, one each in the sixth and eighth years, and the last one in the twelfth year. The exact time of ossification varies among children. Girls often are a little ahead of boys in this process.

The tooth buds of the first teeth begin to form in the human embryo at about four to six weeks and begin to calcify at about 20 weeks. The upper and lower first molars of the permanent teeth begin to calcify very soon after birth. Others begin at three months to three years. The wisdom teeth may not begin to calcify until sometime between the eighth and tenth year, *Food* says.

Naturally, bones and teeth calcify more slowly in

children who have diets deficient in calcium and phosphorus and other essential nutrients. Severe deficiencies can cause permanent stunting of size or malformations of bones and teeth.

Changes occur on both the inside and outside of a bone as it grows larger. New bone is deposited around the outside of the shaft of a long bone. Bone on the inside of the shaft is absorbed at the same time and used elsewhere. Thus the cavity that contains the bone marrow is widened. Bone is added also to the outside of each end of the shaft and then taken from the outside of the area just beneath it.

"Adding material to the outside of bone and subtracting it from the inside gives the skeleton size and strength without unnecessary weight," the yearbook adds. "If bones grew only by adding material to the outside and none were subtracted from the inside, the skeleton would weigh so much that the muscles could not move it around.

"The intricate process of bone building requires many nutrients besides calcium and phosphorus. Vitamin D is essential for absorption from the intestinal tract and the orderly deposition of the bone material. Protein is needed for the framework and for part of every cell and circulating fluid. Vitamin A aids in the deposition of the minerals. Vitamin C is required for the cementing material between the cells and the firmness of the walls of the blood vessels."

Bones can accumulate a reserve supply of calcium and phosphorus at any age if the diet provides enough for the growth and repair and some is left over for storage, *Food* reports.

When the intake is generous, the minerals are stored inside the ends of the bones in long, needlelike crystals, called trabeculae. This reserve can be used in times of stress to meet the body's increased calcium needs if the food does not supply enough.

Of course, when there is no reserve to use, the calcium has to be taken from the bone structure itself—usually first from the spine and pelvic bones. The dentin and enamel of the teeth do not give up their calcium when the body must provide what the diet lacks.

As might be expected, if the calcium that is withdrawn in times of increased need is not replaced, the bone becomes deficient in calcium and subnormal in composition. From 10 to 40 per cent of the normal amount of calcium may be withdrawn from mature bone before the deficiency will show on an X-ray film. Height may be reduced as much as two inches because of fractures of the vertebrae, which are caused by pressure and result in rounding of the back. Such fractures may occur with relatively minor jolts or twists of the body and may not be recognized at the time they happen.

Also, bones with a low content of calcium are weaker and break more easily than bones well stocked with calcium. Breaks in older persons often are related directly to the thinness and brittleness of the bones and are difficult to treat. The bones may be too weak to hold pins or other means of internal repair, and healing may be slow because of the low activity of bone-forming cells, *Food* explains.

"The calcium and phosphorus and other minerals in our food are dissolved as the food is digested," *Food* continues. "Then they are absorbed from the gastrointestinal tract into the blood stream. The blood carries them to the different parts of the body where they are used for growth and upkeep.

"Calcium as it is present in food dissolves best in an acid solution. It begins to dissolve in the gastric juice of the stomach. The calcium is absorbed when the contents of the stomach move into the small intestine. Farther along in the intestine, the contents change from an acid to alkaline reaction, which does not favor the absorption of calcium.

"**Usually 10 to 50 per cent of the calcium eaten is not absorbed but is excreted in the feces.** A small portion of the excreted calcium comes from the intestinal fluids."

"The calcium that is absorbed travels in the blood to places where it is needed, particularly the bones. If any of the absorbed calcium is not needed, it is excreted by the kidneys into the urine. Normal functioning of the kidneys is essential for the normal metabolism of calcium and other minerals," the yearbook says.

As we discuss in another chapter, **vitamin D is essential for the absorption of calcium from the gastrointestinal tract.** Vitamin D does not occur naturally in many foods. Egg yolk, butter, fortified margarine and certain fish oils are the chief sources. To a few foods, notably milk and cereals, some vitamin D is added.

Food reports that a special substance, cholesterol, is present in the skin and is changed to vitamin D by the ultraviolet rays of the sun. We cannot be sure that enough of the rays reach the skin and produce vitamin D in all seasons of the year and in all parts of the country to insure normal growth. Most infants and young children, therefore, are given daily a concentrated source of vitamin D, such as cod-liver oil or some other fish oil. Adults normally do not need more vitamin D than they customarily get from food and exposure of the skin to the effective rays of sunshine.

Adults who stay out of the sun entirely or wear clothing that covers all of the body except the face probably do not get enough vitamin D naturally and, therefore, need to have some added to their diet. This would especially apply to people who live in northern climates, where the sun is not so bright in the winter months.

Food points out that too much vitamin D can be dangerous. It overloads the blood and tissues with calcium. Infants who are given several times the amount of vitamin D that they actually need may suffer gastrointestinal upsets and retarded growth. This condition is called hypercalcemia—meaning too much calcium in the blood. It can be cured, if it is recognized soon enough, by simply omitting the vitamin D from the diet.

A hormone secreted by the parathyroid glands has an important part in the body's use of calcium and indirectly in the use of phosphorus, we are told by *Food.* There are two of these tiny glands on each side of the neck near or in the thyroid gland.

The parathyroid hormone keeps the amount of calcium in the blood at a normal level of about 10 milligrams per 100 milliliters of blood serum. (Serum is the watery part of the blood that separates from a clot.)

Any wide deviation from this amount is dangerous to health and life, *Food* says. The hormone can shift calcium and phosphorus from the bone into the blood. If the blood levels are too high, it can increase the excretion of these minerals by the kidneys. If anything reduces the secretion of the parathyroid hormone, the calcium in the blood drops quickly, the phosphorus rises, and severe muscular twitchings result.

As expected, the body will absorb more calcium when it is needed for growth and for storage during pregnancy and lactation (the period after birth when a mother produces milk for breast-feeding), or after periods of loss than it will when such needs do not exist.

"The body is likely to absorb a larger proportion of calcium from a low intake than from a generous intake," *Food* says. "The body tries in this way to make full use of a small supply, especially when its needs are great. In terms of total amount, however, more calcium is absorbed from a generous intake than from a low one."

As discussed in other chapters, lactose, the form of sugar present in milk, is especially good in promoting the absorption of calcium. Certain proteins and amino acids also are effective. Perhaps the combination of these is responsible for the excellent absorption of calcium from milk.

The absorption of calcium from vegetables is somewhat lower, *Food* says. The high content of fiber, especially in the coarse, leafy, green vegetables, makes them move through the intestine rapidly, and the amount of calcium absorbed is reduced.

Spinach, beet greens, chard and rhubarb contain a chemical, oxalic acid, which combines with the calcium to make calcium oxalate, *Food* continues. Because it is insoluble in the intestinal fluids, the calcium cannot be absorbed but is excreted in the feces. We should add that these are excellent foods, nutritionally speaking, but you should not eat them every day.

Food also reports that the outer husks of cereal seeds, such as wheat, contain phytic acid, a substance that combines with phosphorus to form phytates. The phytates can interfere with the absorption of calcium, especially in a child when a high intake of phytic acid is accompanied by an inadequate supply of calcium and vitamin D. Phytates are not likely to hinder the absorption of calcium in the diets commonly used in the United States, *Food* adds.

It should also be noted that laxatives are also likely to lower the absorption of calcium. In *Let's Get Well*, Adelle Davis reports that laxatives and cathartics irritate delicate intestinal membranes and interfere with digestion and absorption of essential nutrients. Mineral oil, probably the most damaging of all laxatives, she says, decreases the absorption of calcium and phosphorus and itself absorbs the fat-soluble vitamins—A, D, E and K—and carotene from the foods in the intestine; it passes into the lymph and blood, picks up more fat-soluble vitamins from liquids and tissues throughout the body, and is then excreted in the feces.

A University of Rochester scientist has discovered, he says, that **calcium in the diet reduces tooth decay**. He could, of course, have discovered this fact by reading some of the health food literature of the past 50 years or so. But, no. He fed animals tooth-decay causing foods—chiefly snacks and commercial cereals loaded with sugar. And he produced rampant tooth decay in the animals.

Then he added calcium lactate to the sugar-laden foods and produced a 50 per cent reduction in tooth decay. The doctor says there are no known bad effects connected with "the drug" as scientists prefer to call nutrients when they give them to prevent some illness.

Why, said the scientist, we could add this calcium to all these sugar-laden foods without destroying the taste of them and thus could cut in half incidence of tooth decay in children. Just what the health food movement has said for many years when they recommended bone meal, dolomite, calcium lactate or any other calcium preparation to prevent tooth decay.

Of course, at the same time we do not recommend the sugar-containing snacks and cereals. It seems never to occur to a scientist working in the field of tooth health that something which is destructive to teeth is also probably equally destructive to other parts of the body. As is becoming increasingly evident these days, eating sugar in the amounts in which we eat it is the single most destructive eating habit we have.

We usually think of calcium only in terms of teeth and bones, forgetting that **a considerable part of the calcium in our bodies is needed in the blood to help control the heart beat and nourish the heart.** An article in *Medical World News* for July 12,1976 deplores the fact that many patients who receive rapid blood transfusions or post-operative hemodialysis may risk serious heart problems because there is just not enough calcium in the blood being transfused.

Citrate is used to preserve stored blood The calcium which healthy blood contains normally is bound by the citrate and hence not available for regulating the heart beat. Experiments described in the article revealed that in several cases of blood transfusions calcium was given simultaneously and blood pressure rose, as desired, 18 per cent. Cardiac output increased 51 per cent. In other experiments without added calcium cardiac output increased only 26 per cent and blood pressure rose only 10 per cent.

It seems that research on this subject has been going on since 1951 and presumably the medical profession has known all this time what the effects of calcium or lack of calcium will be. Why then do they not routinely add this mineral to transfusions? The doctors who report on these incidents are not sure just why. But it seems that medical textbooks are vague on the subject.

A high level of calcium in the diet protects to some

extent against the possibility of lead poisoning, according to a news release from Cornell University in August, 1973. Eighteen tons of lead are deposited daily in the air over Los Angeles, the city of automobiles. Most of it apparently accumulates in soil, water, and, of course—people. The news came from the American Chemical Society in August, 1973.

After a series of tests called the most extensive and conclusive ever conducted in space, NASA scientists have announced that human beings can apparently live healthfully in a condition of weightlessness *except*.... And it's an important exception. Loss of calcium from bones on long space flights may be so severe that astronauts may break backs or legs or may develop kidney stones because of accumulations of the lost calcium in the kidneys. Dr. Leo Lutwak of UCLA reported that Skylab 3 astronauts lost calcium throughout their 84-day mission. No one knows whether such losses will be reversed after more time in space or whether they will continue. Nor do the experts know from where the calcium was lost. If it was lost chiefly from leg bones, they believe that exercises may prevent permanent damage. If from the spine, then astronauts may have to take hormone drugs to prevent this loss. Skylab astronauts lost calcium from their heel bones but not from their arms, we are told.

Still another warning on the dangers of a low-calcium diet came from a Mayo Clinic researcher in April 1976. Dr. Jenifer Jowsey, director of orthopedic research at the famed clinic, told an audience at the Sixth Annual California Dairy Council meeting that the average American diet, high in phosphorus and low in calcium almost guarantees osteoporosis in later life.

Said Dr. Jowsey, "Americans in general tend to decrease their intake of calcium as they get older. However, the shift from big meals to snack foods has caused phosphorus intake to go up and calcium intake to go down in the United States." Foods which contain a great deal of phosphorus and little calcium are: fish, poultry, meat, cereals and bread. Dairy products, too, are rich in phosphorus, although they are also our best source of calcium.

All these are fine foods—no reason to slight them. But, says Dr. Jowsey, **if you want to avoid the painful and disfiguring disease of osteoporosis in later life, you must somehow manage to get more calcium.** She suggests taking calcium supplements, not just after middle-age when you become concerned about the dangers of softened, brittle bones, but from the age of 25 on. Take supplemental calcium up to one gram daily in the form of calcium tablets, she says, if you want to avoid osteoporosis in later life. If you haven't done this earlier in life, of course, the next best thing is to begin now to use calcium supplements daily.

The official recommendation for calcium intake is 800 milligrams a day for adults. Dr. Jowsey believes we need more than that and a number of prominent researchers in this field agree with her. They, too, suggest 1,000 milligrams daily in a supplement—that is, in addition to the amount in food.

Dr. Jowsey says it is essential for life that the level of calcium in the blood remains normal. "When there is an abnormal absorption of this element in the stomach and intestines or when too much calcium is lost in the kidney, the skeleton provides the only source of calcium. Bone tissue is resorbed (taken away, borrowed) to put more calcium in the blood." The same thing happens when the diet is too high in phosphorus without enough calcium to balance it.

"If you want to end up at 70 with the skeleton of a 16-year-old, start supplementing your diet with calcium at age 25," says Dr. Jowsey.

She mentions two other aspects of life which are related to the way our bodies use calcium. Alcoholic drinks tend to make us lose calcium. Half of all the calcium we get in food may be lost if we drink something alcoholic at the same time. The other important thing to keep in mind is activity. Daily activity is absolutely essential to keep bones healthy.

Inactivity (very common in older women) prevents stress to the bones. And they need stress . . . the stress of holding the body upright, of supporting the legs, hips and arms during strenuous activity. Just going for a 15-minute walk or taking 20 minutes of exercise is not enough, says Dr. Jowsey. "One needs to be active throughout the day for the proper stress effects to occur."

"Physical fitness" means endurance, according to Dr. Alexander Leaf, a prominent gerontologist (student of old age), who writes often and entertainingly about the value of consistent daily exercise on health. Dr. Leaf has journeyed to three of those pockets of longevity and good health where modern medical men are astounded to find people over 100 years of age who are still working in fields and orchards, still doing all their own garden and house work, riding horses, dancing, climbing high hills and mountainsides with the agility of mountain goats.

In the August, 1977 issue of *Executive Health*, Dr. Leaf adds another brilliant study of longevity and fitness to his earlier writings on this subject. He describes his visit to Soviet Georgia where the famous yogurt-eaters live to a great old age in good health—active, sprightly and hard-working up to their very last moments—and apparently in excellent mental health as well. We can find not any mention of what we call senility among the accounts of these long-lived people.

Dr. Leaf tells us that broken bones among the elderly Georgians are almost unknown for their bones do not succumb to osteoporosis as those of our old folks do. The chief reason, says Dr. Leaf, is the physical fitness of these people, attained through constant vigorous physical activity which strengthens bones and prevents loss of minerals which occurs during inactivity. He reminds us that even our young astronauts suffered grievous loss of minerals during space trips because of lack of "stress" on bony structures, as we have just noted.

"Exercise should be regular, frequent and continued throughout life," Dr. Leaf reports. "Endurance exercises are most beneficial—not speed or strength, but endurance is what counts. . . . Long walks, jogging, cross-country skiing, swimming, rowing and bicycle riding are the recommended sports . . . at least throughout your middle years. These are activities which can be paced, so as not to be strenuous, and nearly everyone can participate provided that he starts modestly and increases activity gradually."

He noted that "These men I have examined around the world who live in vigorous health to 100 or more years are great walkers. If you want to live a long, long time in sturdy health you can't go wrong in forming the habit of long vigorous walking every day . . . until it becomes a habit as important to you as eating or sleeping."

Hip and pelvic bones need to bear weight, to support us as we stand erect and move about. Arm bones need to bear weight as we lift and carry heavy things. The back needs a certain amount of stress in order to keep all its complicated interwoven patterns of bones functioning at their best. **With inactivity at any age, says Dr. Leaf, our bones lose their calcium salts and become thin and fragile—another excellent reason for maintaining such superlative good health that you will never be confined to bed for even a day's illness.**

He adds that, in spite of efforts with hormones, vitamin D and high intakes of calcium and phosphates, the treatment of osteoporosis in our old people has resisted cure. Continued physical exertion remains the most potent preventive measure against this disability, he says.

It's not just bones which benefit. The heart and entire circulatory system benefit to such an extent that researchers have found greatly superior heart and circulatory health in people who walk for even one to nine minutes as compared to those who do no walking at all. In a British study, those who walked to work for 20 minutes had better circulatory health than those who walked less. And the longer the daily walk, the greater the improvement in heart and circulation.

Earlier in this chapter we reported on the potential dangers of eating sugar. Dr. Marshall Ringsdorf of the University of Alabama said in a recent discussion of the role of sugar in our diets that there is evidence that **sugar increases the excretion of some nutrients to produce secondary deficiencies. That is, just eating sugar causes your body to lose other essential nutrients—minerals, for example.**

Larger amounts of calcium are found in the urine after you eat sugar than after you eat most other foods. This is especially true, he says, of people who have a personal or family history of kidney or bladder stones. "There is a significant drain on nutrient supplies when glucose comes into the blood stream from sucrose (sugar)," says Dr. Ringsdorf. The sugar brings along no nutrients (no protein, vitamins or minerals) but requires nutrients in order to be used by the body.

For at least two hours after one has eaten two or four ounces of sugar, the phosphorus in your blood decreases from about 3½ or 4 milligrams per cent down to about 1½ milligrams per cent. This means that for about two hours after this much sugar is eaten, there is often so little phosphorus in the blood that calcification or bone formation cannot take place. Bone is constantly being replaced and reformed so in people whose sugar consumption creates a negative balance of phosphorus, destruction of bone may exceed bone formation for two to five hours every day. "This is part of the problem of osteoporosis, both in the mouth and in the rest of the body," says Dr. Ringsdorf.

It's interesting to remember, in reference to these findings, that our federal government subsidizes the sugar industry to the tune of some 240 million dollars a year of taxpayers' money. So while one branch of the federal government is spending our money trying to find "cures" for diseases caused by excessive sugar consumption, another branch is giving subsidies to people who raise sugar, so that these diseases will eventually become even more widespread and devastating.

Recent work at the University of Texas has shown that **calcium may be extremely important for the health of heart muscles.** It influences the contraction and relaxation of these muscles. The amount of exercise you get also influences the effect of calcium on the heart muscles to such an extent, say these Texas researchers, that animals which are exercised vigorously survive heart attacks much better than those which get no exercise. And the effect carries over into the next generation, for offspring of these healthy animals have much stronger hearts.

As we have indicated, **you lose calcium when you are lying on your back for long periods of time.** The only thing that will correct this and return the body to a normal calcium balance is standing.

A physician at a Philadelphia hospital experimented with a number of healthy young volunteers. They were kept in bed lying on their backs for considerable lengths of time. They lost large amounts of calcium in their urine. They were given vigorous exercises which they could take lying on their backs. This did not change the calcium picture. They were given differing amounts of calcium in their diet. This did not change the calcium excretion. The bed exercises were increased up to three or four hours daily. There was no effect.

But as soon as the young men were gotten out of bed and required to stand quietly for two or three hours daily, their calcium balance returned to normal. The doctor in charge concluded that it is the action of gravity, rather than anything else which keeps us from losing calcium when we are spending some of our time standing. This finding would seem to be extremely important for many groups of people. Our astronauts, for example. And, of course, hospital patients.

A Cornell University professor of psychiatry has discovered that the body's use of calcium is upset by various curative procedures used by psychiatrists. He studied patients in a mental hospital, all of whom were getting the same amount of calcium in their diets and all of whom were being treated with electric shock or a tranquilizer. **He discovered that some of the patients with certain kinds of mental disorders lost calcium in their urine, consistently, during treatment.** Those with other disorders did not. And, significantly, the patients who continued to lose calcium during the treatment were those who improved. The ones who did not improve did not show a difference in the way their bodies used calcium.

The author asks if it might be possible that the loss of calcium had something to do with the patients' ability and opportunity to move around. And he came to the conclusion that, because of the excellent therapy being given these particular patients, their mobility or lack of it could not have influenced the results. So he concludes that calcium is certainly involved very closely in the improvement or lack of improvement of patients with certain kinds of mental illness. This seems to suggest to those of us who are healthy and want to remain that way that we guard very carefully our sources of calcium and make certain that we get plenty of it every single day. This is especially important in children who need lots of minerals to help them grow bones.

CHAPTER 2

Everyone Needs Calcium-Rich Foods

TODAY THERE IS much confusion and lack of knowledge about many of the trace minerals. Although we are gradually uncovering valuable facts about their place in nutrition, we still do not have enough information to recommend amounts of all of them which we should be getting in food, or to know very much about the possible toxicity of some of them.

As we are discovering in this book, calcium is a different story. As has its companion mineral, phosphorus, calcium has been well researched for many years. Physicians have known of its benefits since the early days of the herb doctors. And today our nutrition specialists believe they know most of the functions of this essential mineral in our bodies. The one thing everybody is agreed on is that you can't get along without it. In fact, you can't get along at all well, if you are even a bit deficient in it. And some experts are suggesting that we may need 1,200 milligrams or more a day, especially as we grow older. This is far above the recommended daily dietary allowance.

How many times have you heard middle-aged people say they never drink milk because milk is a food for children only? "Children must have the calcium of milk for forming bones and teeth," say such people, "but you don't need much calcium as you grow older."

How wrong such sentiments are and what nutritional trouble they can get you into are pictured graphically by an assistant director of the Iowa Agricultural Experiment Station in an article in the March, 1955 issue of *Iowa Farm Service.* Dr. Pearl Swanson and a collaborator from the United States Department of Agriculture title their article, "You Don't Outgrow Your Need for Calcium."

They describe a survey made among more than 1,000 Iowa women to discover just how much calcium they were getting in their everyday meals. All the women were 30 years old or older. They were a representative cross-section of the population, so they actually represent about half a million Iowa women who eat as they do. **The experts who conducted the survey found that only one woman in every five was getting in her meals and snacks the amount of calcium recommended officially as the best amount.** The reason was simple. They didn't drink enough milk or use enough foods that contain milk, like cheese. Although they were eating fairly good diets from the point of view of other nutrients, their supply of calcium was so low that they could expect, as time went on, to be bothered by many kinds of disorders that result from lack of calcium.

As we know, bones and teeth continue to need calcium and other minerals all during our lives, for minerals are constantly being lost and have to be replaced. In addition, calcium is needed for the muscles to relax and contract. This happens every time we move. Since the heart is a muscle, enough calcium is essential for regulating the heart beat. It is also involved in the proper clotting of blood. Without enough calcium in the blood, you might bleed to death from a minor cut.

"Middle-aged bones seem to break more easily than young ones," says Dr. Swanson, "and we think this occurs largely because calcium has been withdrawn and not replaced. Our hipbones, for instance, carry much of the body weight. If the calcium in our diets is inadequate, the hipbones may become so weak that they no longer are able to support this weight. Thus, a bone may break and we fall."

What did the Iowa women eat that gave them fairly adequate supplies of other nutrients, but not nearly enough calcium? Here is a list of a typical day's menu. It includes: a serving of meat, fish or poultry, an egg, several slices of bread, a serving of white potatoes, a serving of corn, an orange or some orange juice, a serving of tomatoes and one of peaches, a salad of lettuce, cucumbers, radishes and onions, a little cream, some butter and salad dressing, a piece of cake, some sugar and jelly. Does this sound like the kind of food most of your family and friends eat?

Such a diet contains about 300 milligrams of calcium. The recommended daily allowance of calcium is 800 milligrams—

How Much Calcium Do You Need?

According to the National Academy of Sciences, we need calcium every day in the following amounts. We give the amounts in milligrams (or mgs).

Men and Women need	800 mgs.
Pregnant and lactating women need	1,200 mgs.
Infants need	360-540 mgs.
Children 1-10 years old need	800 mgs.
Boys 11-14 need	1,200 mgs.
15-18 need	1,200 mgs.
Girls 11-14 need	1,200 mgs.
15-18 need	1,200 mgs.

more than twice what these women got in their daily food. "Such a diet cannot be regarded as satisfactory," says Dr. Swanson. By adding two cups of milk they could bring the calcium up to the recommended amount. The milk can, of course, be taken as a beverage, or used in cooking (sauces, puddings, casseroles, etc.), it may be eaten on cereal or taken in cheese or yogurt. One ounce of cheddar cheese supplies about as much calcium as three-fourths of a cup of milk.

But the point is you must somehow manage to get a full two cups of milk or its equivalent every day. **You can't substitute anything for calcium. And without milk it is almost impossible to get the recommended daily amount.** Can't you get calcium from fresh leafy vegetables?

Calcium Content of Foods

Almonds	234 milligrams in ¼ pound
Bread, whole wheat made with dried milk	118 milligrams in 3 slices
Buttermilk	121 milligrams in ½ cup
Carob flour	352 milligrams in ¼ pound
Cheese, cheddar	750 milligrams in ¼ pound
Cottage	94 milligrams in ¼ pound
Swiss	925 milligrams in ¼ pound
Collards	250 milligrams in ¼ pound
Dandelion greens	187 milligrams in ¼ pound
Filberts	209 milligrams in ¼ pound
Kale	249 milligrams in ¼ pound
Milk, whole	576 milligrams in 2 cups
Powdered, skim	1,308 milligrams in ¼ pound
Mustard greens	183 milligrams in ¼ pound
Sesame seed (whole)	1,600 milligrams in ¼ pound
Seaweed, kelp	1,093 milligrams in ¼ pound
Soybean flour, defatted	265 milligrams in ¼ pound
Whey, dried	646 milligrams in ¼ pound
Yogurt	588 milligrams in 2 cups

Yes, you can, but you must eat large amounts of these every day.

For instance, you would have to eat three pounds of cabbage or two pounds of endive, or one pound of cooked kale, or two pounds of lettuce or one pound of mustard greens to get the 800 milligrams of calcium which is contained in three cups of milk. In addition, there is the problem of absorption. Some leafy vegetables contain substances that make the absorption of calcium difficult. But milk digests and is absorbed perfectly in all normal, healthy people.

What about yogurt? What about goat milk? Yogurt and goat milk, cup for cup, contain just the same amount of calcium as cow's milk. In addition, yogurt contains the helpful lactobacillus bacteria which are good for digestion and the health of the digestive tract.

Goat milk has the added advantage of being naturally homogenized. That is, the fat and non-fat parts of the milk are naturally mixed in such a way that they do not separate as they do in cow's milk. Some people find this milk much easier to digest because of this natural homogenization.

Then, too, goats are generally raised by health conscious people who have small dairies and take great pains to feed and care for their animals.

There are other excellent sources of calcium available at your health food store. Sesame seed, alone among all the seed products, is quite rich in calcium. One-fourth of a pound of whole sesame seed contains more than 1,000 milligrams of the mineral. And there are many calcium food supplements available, all pleasant-tasting and easy to take.

If you are allergic to milk or for some reason cannot use it, the calcium supplement becomes an absolute essential. And even if you do use lots of milk, the added calcium in a supplement will be beneficial. There is little danger of your getting too much calcium if the rest of your diet is well balanced. Yogurt and cultures for making yogurt are available at your health food store. Many of the complete food supplements contain such calcium-rich milk products as whey.

Why not check through your day's eating and that of your family, for a week or so, just to make certain everyone is getting enough calcium. Pay special attention to the very young and the older members of your family. Older people tend to neglect calcium-rich foods, and you may save them much distress from softening bones and serious falls later on, if you insist on plenty of calcium-rich foods in their diets now.

In the following chapters, we discuss milk, yogurt, cheese, etc., in greater detail.

CHAPTER 3

How Much Phosphorus Do You Need?

WHEN WE SPEAK of many minerals, we are talking about elements that are needed in extremely small amounts—just smidgeons of this or that which are essential. But in the case of phosphorus, the wholesome, healthful diet must give you fairly large amounts of this mineral. It is needed for so many functions in your body that it becomes more than just something you can afford to slight, nutritionally speaking.

You need the phosphorus; you could not live without it. And, yet, because of the way we modern Americans eat, phosphorus plays its important part in our diets mainly because of its relationship with calcium. It is generally agreed among nutritionists that we are in no danger of getting too little phosphorus in these times. Why, then, should we bother about it at all?

Meat, fish, eggs, dairy products, cereals and nuts are all rich in phosphorus. These foods make up a considerable part of our meals, so it is unlikely that any of us lack phosphorus. But all of these foods—except milk—tend to be rather low in calcium. And that is where the difficulty arises. As we know, the more phosphorus you get, the more calcium you need, for these two minerals are inexorably linked in such a way that you cannot neglect one and expect the other to be used by the body to perform all the functions it is supposed to perform.

Let's say you plan to eat the very best possible diet, from the point of view of protein and vitamins. And even food supplements. You eat lots of meat, fish and poultry. You add wheat germ and brewers yeast, excellent sources of B vitamins and minerals. You eat eggs and nuts for their protein, their vitamins and their wholesome fats. But, for one reason or another, you decide not to add any milk or dairy products. You have designed a diet rich in almost every food element—except calcium.

Such a diet, because of its high phosphorus content, makes your need for calcium even greater. So, although you are eating all these nourishing foods, you might end disastrously and suffer very seriously from disorders involving calcium deficiency. Leafy vegetables contain considerable amounts of calcium relative to other foods. But even if you added these to the diet outlined above, it would be extremely hard to eat enough of them to balance the large amounts of phosphorus in the meat-cereal-seed-egg-rich diet.

There seems to be a sound nutritional reason why "bread and cheese," "bread and milk," or "oatmeal and milk" are established partners in diet in many parts of the world. Mexicans who live mostly on tortillas made from ground corn (rich in phosphorus) add limestone (rich in calcium) to the corn when they grind it. Centuries of observation have shown these people who have certainly never studied nutrition that they need a source of lime or calcium when they eat large amounts of cereal or seed foods.

Phosphorus is present in every tissue of our bodies and performs more functions than any other mineral. About 80 per cent of the body's phosphorus is in bones, where it is combined with calcium to form the skeleton, the bony framework of our bodies. It is present in every cell; it participates in the working of muscles (including the heart), it aids the body in using carbohydrates, protein and fat in food, it is involved with the health of the nerves, the chemical changes that normally take place in blood. In addition, it is part of many "enzyme systems." That is, it works along with proteins, vitamins and enzymes to bring about certain necessary biochemical functions of the body.

Calcium and Phosphorus Content of Some Common Foods

(Remember, you should aim at about 1-1/2 times as much phosphorus as calcium)

Food	Calcium in 1 serving (Mg. or milligrams)	Phosphorus in 1 serving (Mg. or milligrams)
Whole liquid milk	288 mg. in 1 cup	230 mg. in 1 cup
Milk, powdered, skim	520 mg. in ½ cup	850 mg. in ½ cup
American cheese	133 mg. in 1-inch cube	130 mg. in 1-inch cube
Yogurt	294 mg. in 1 cup	230 mg. in 1 cup
1 egg, whole	27 mg.	112 mg.
1 serving lean beef	10 mg.	214 mg.
1 serving chicken	10 mg.	232 mg.
1 serving haddock	15 mg.	197 mg.
10 almonds	25 mg.	45 mg.
20 cashew nuts	16 mg.	160 mg.
18 peanuts	15 mg.	73 mg.
1 serving peas	28 mg.	127 mg.
1 serving potatoes	9 mg.	52 mg.
Whole grain bread	20 mg. in 1 slice	102 mg. in 1 slice
Wheat germ	70 mg. in ½ cup	1050 mg. in ½ cup
Brewers' yeast	49 mg. in 1 heaping tablespoon	945 mg. in 1 heaping tablespoon

The National Research Council, official government body for decisions on scientific things, recommends that we get in our food about one-and-one-half times more phosphorus than calcium. **They do not set any minimum requirement for phosphorus, for they believe that we are all getting plenty of it.** They recommend that infants, pregnant women and nursing mothers should get not more phosphorus—but less—in relation to the calcium in their diets.

That is, these people whose need for calcium is greater than anyone else's should get the same amount of phosphorus as calcium. They should make an effort to get even more calcium than the rest of us do, in relation to the amount of phosphorus they get. As you can see, when we talk about phosphorus, we always speak relatively—phosphorus requirements are always relative to calcium requirements.

This, then, is the basic thing to remember about phosphorus when you are planning meals. **It is an important mineral and it exists in ample quantity in many valuable foods.** See the chart for the phosphorus and calcium content of some common foods. Note that there are very few in which the calcium content comes anywhere near the phosphorus content.

The more of these excellent phosphorus-rich foods you eat, the larger your requirement for calcium becomes. Wheat germ, for instance, is one of our best foods. It contains, gram for gram, more phosphorus than any other food except brewers yeast and not a great deal of calcium, relatively speaking. Does this mean that you should not eat wheat germ or brewers yeast? Not at all. It means that you should be careful to include plenty of calcium in your meals at all times. If you love wheat germ, if you eat a lot of it and add it to many other foods to enrich them, or if you are eating large amounts of brewers yeast, then you need more calcium to balance the phosphorus you are getting.

Powdered milk contains about 13 times as much calcium as wheat germ. So you would do well to add powdered milk to foods containing wheat germ and brewers yeast and eat plenty of milk when you use wheat germ as a breakfast cereal. Cheese is another food rich in phosphorus and calcium. An excellent addition to any meal.

Foods Rich in Phosphorus

(The figures are given in terms of one average serving—about 1/4 pound)

Food	Milligrams of Phosphorus
Almonds	504
Beans, kidney	406
Beans, mung	340
Beef	200
Bran flakes	495
Chicken	265
Chickpeas	331
Cowpeas	426
Eggs, 2	200
Filberts	337
Flounder	344
Lentils	377
Liver	476
Peanuts	407
Peanut flour	720
Pine nuts	500
Pumpkin seed	1,144
Rice bran	1,386
Rice polish	1,106
Safflower seed meal	620
Sesame seeds	616
Soybeans	554
Soybean flour, defatted	655
Sunflower seed	837
Sunflower seed flour	898
Wheat bran	1,276
Wheat germ	1,118
Yeast, brewer's, 1 tablespoon	945

The relationship of calcium and phosphorus is one of the best illustrations of the necessity for eating a widely varied diet, for these two minerals occur in entirely different kinds of food and, if your diet is one-sided, you are likely to be short on one or the other. Calcium occurs in the leafy, fresh parts of plants, chiefly stalks and leaves. Phosphorus occurs chiefly in the seed part of the plant. So if you try to live on salads only, you may invite a shortage of phosphorus. If you try to live on the seed part of the plant—that is, beans, cereals, nuts, etc.—you may become deficient in calcium.

Phosphorus is absorbed most effectively when the strongly acid digestive juices are present in the stomach and intestine. As we grow older, these digestive juices tend to decrease, so the older person may actually need more phosphorus (and calcium) than the younger person, just because he may absorb less from his food. So the older person who shuns seeds and nuts, peanut butter, nut butters, meat, eggs, etc., may lack phosphorus. If he also dislikes milk, he is almost certain to be courting a calcium deficiency.

As we stated, wheat germ has almost as much phosphorus as yeast. And wheat germ, of course, is eaten as a cereal in servings as large as any other food. Sesame seeds and sunflower seeds are extremely rich in phosphorus. Dry soybeans contain more phosphorus than meat. Since soybeans are the nearest thing to a complete vegetable food, it is not surprising to find that they are also a rich source of calcium. Soybean flour is unmatched as a source of both calcium and phosphorus.

And don't forget that all nuts are good sources of phosphorus. We recommend eating them without salt or roasting, just as they come from the shell. And don't forget that other delicious snack food—seeds. All are rich in phosphorus, as well as protein, healthful fats and iron.

Have you often wondered why your health food store has some brands of phosphorus-free supplements, while other brands stress the fact that their product contains both phosphorus and calcium, as in the case with products like bone meal?

The reason is probably by now obvious. In most American diets there is plenty of phosphorus and probably not enough calcium. Of course, we are making a mistake by even saying such a thing, since, obviously, such a concept as "most American diets" is an error. The only thing that should concern you is your own diet and that of your family. We cannot talk in terms of your diet, since we do not know what you eat every day.

To put it another way, **if you are one of those Americans who gets lots and lots of phosphorus in your daily diet and probably not enough calcium, then you will probably want to take a supplement of calcium which contains no phosphorus—you are already getting plenty of phosphorus—right?** If, on the other hand, you make a real effort to get plenty of the high-calcium foods every day and feel certain you are getting enough calcium, then you may wish to take a mineral supplement which contains both these minerals, plus iron, magnesium, copper, etc.

If you seldom eat nuts, beans, seeds, soybeans, wheat germ, brewers yeast, rice bran or rice polish, eggs, etc., and feel that you may not be getting sufficient phosphorus, then you might wish to take the calcium-phosphorus supplement.

CHAPTER 4

Vitamin D and Calcium Are Essential for Strong Bones

NO MATTER HOW often reports of devastating calcium deficiency among older folks appear in medical and scientific journals, we read new reports all the time. Doctors continue to be surprised and shocked at the condition of the blood and bones of older folks where calcium is concerned. Another lack which often accompanies the calcium deficiency is lack of vitamin D, as we have seen in several other chapters in this book.

Doctors have known for about 70 years that lack of vitamin D means lack of calcium in blood and bones. This is because the vitamin is absolutely essential for the intestines to absorb the mineral calcium. And phosphorus. It's not often that anybody suffers from lack of phosphorus, since there is plenty of this mineral in meat, eggs and all cereal products.

But calcium is another matter. Unless the individual is very consciously striving to get enough calcium at meals and in supplements, it's very easy to neglect it. Calcium and phosphorus balance one another in body chemistry, so the more phosphorus you get the more calcium you may need. A diet with lots of meat, eggs, bread and cereals is quite deficient in calcium. And unless it is buttressed with plenty of dairy products it can bring great grief and suffering, especially to older folks.

In *Geriatrics* for January, 1977, three British physicians describe their findings at a hospital in England where 42 per cent of all the older people admitted to the geriatric unit were suffering from too little calcium in their blood. As age went up so did the deficiency. **The older men and women became, the less calcium they had in blood and bones.** The result was a condition called *osteomalacia*, which is adult rickets.

This is how it works. Vitamin D is lacking, so little or no calcium is absorbed from the digestive tract even if plenty of calcium is being eaten. The levels of calcium in the blood must fall. This stimulates the parathyroid glands and they secrete more of their hormone. The body is desperately trying to attain a normal blood level of calcium. The parathyroid hormones release calcium from bones into the blood and sooner or later bones become so frail that they fracture.

As like as not, neither the older person nor the doctor knows that this process is happening until great destruction of bone has taken place, especially the hip bone which is more likely to be the source of painful falls among the elderly.

A mild deficiency in calcium, caused either by lack of calcium in food, lack of vitamin D, or both, can show itself first by lethargy, apathy, vague, generalized pains. The patient tires easily. He or she may suffer from low back pain. Then both muscles and bones begin to hurt. "At this stage," say the English doctors, "rheumatoid arthritis, fibrositis or lumbosacral strain is sometimes diagnosed and the patient is then treated accordingly without much improvement."

The doctor may decide the patient is neurotic, for there

may be depression and irritability and sometimes much more startling symptoms like hallucinations. The patient thinks he may just be getting old and so should expect these conditions. **But as bones begin to hurt more and more, as the patient begins to walk with a waddling gait and bone fractures appear, the doctor may decide finally that this is adult rickets or osteomalacia.**

The English doctors have devised a test which, they say, will give the doctor a very good idea whether his patient has osteomalacia so that he can treat it long before any serious symptoms develop. But why should we be concerned with such a test? What is the matter with simply finding out whether the patient has been getting enough calcium in food and enough vitamin D from food or from sunlight? If not—and it's more than likely he has not—then for goodness sake let's prescribe calcium supplements and fish liver oil for vitamin D! And give the patient a diet that will cure the calcium deficiency!

We suspect this seldom happens because nutritional deficiency is apparently the last thing most doctors ever suspect when they diagnose disease, even though medical journals are brimming over with accounts of gross deficiencies in protein, minerals and vitamins in many people who come to doctors. All they need is a change in their meals so that more protein, minerals and vitamins are available to them. But doctors seldom think of this.

Vitamin D is manufactured on the bare skin when it is exposed to sunlight. It should be obvious to any physician that any old person has less opportunity to get out in the sunlight, especially in winter, than young people have. In northern countries there is much less sunshine in winter. The winter cold necessitates bundling up in heavy clothing so that almost no sunlight gets to the skin, where it can manufacture vitamin D. So why not issue a blanket recommendation that everybody past middle age take a vitamin D supplement, especially in winter?

Many older folks avoid dairy products because they're

relatively expensive, they cannot be stored for very long, and many people think they are suitable foods for young people only. It's not so. Milk, cheese of all kinds and yogurt are the only reliable sources of calcium we have. It's almost impossible to get enough of this essential mineral without eating these foods every day. And this shouldn't mean a few teaspoons of milk in coffee or a thin slice of cheese on top of your sandwich. Drink at least two glasses of milk every day or the equivalent in cheese or yogurt. Use milk and cheese in cooking. Make soups of milk and vegetables. Use cheese in casseroles. Use it for dessert along with some fruit, dried fruit or nuts.

Several years ago we ran across a note in *Science News* indicating that **diabetics may have many more bone troubles than healthy people because they have a problem with using vitamin D.** Three chemists reported on experiments with diabetic rats showing that the rats were able to absorb vitamin D from their food but were apparently unable to break it down into the normal body compounds after absorption. These compounds are essential for binding the mineral to a protein so that it can be incorporated into bone structure. When the scientists injected vitamin D into their diabetic rats, they had no trouble absorbing it and changing it into the proper body substance.

"The effects of diabetes on normal body functions are more far reaching than we hitherto expected," said one of the scientists, adding that there is no guarantee that the calcium disorder is present in diabetic human beings as in rats, but it seems likely that it is. This is still another reason for us to eat the best possible diet and concentrate on getting enough vitamins, minerals and protein so that we will not invite diabetes and, if we are diabetic, we will perhaps avoid complications like this.

A new kind of vitamin D was recently developed by a team of California scientists who hoped they might be able to prevent complications of uremic patients. It seems that patients with kidney diseases have trouble with bones which is so extreme that, in 34 of the patients described in an article in *Medical Tribune,* severe symptoms including bone pain and fractures, muscular weakness and lack of calcium in the blood, as well as abnormal X-rays were reported. Other patients on kidney dialysis had less severe symptoms but definitely lacked calcium and had abnormal bone biopsies.

These patients had been treated with up to six grams of calcium daily in an effort to get their body calcium back to normal. No success. So the newly developed kind of vitamin D was used along with calcium. The reason the doctors expected success was that much more of the new vitamin D can be given in this form than in the "natural" form of fish liver oil, which becomes toxic if too much is given.

In more than two-thirds of the patients who had skeletal pain, the pain began to subside within several weeks and disappeared entirely in more than half of the patients. Muscular weakness improved in a number of them. The amount of calcium in the blood of all patients increased, demonstrating that the new vitamin D was indeed doing its assigned job of helping the bodies of these very ill people to absorb and use calcium as they were meant to.

These, then, are the two closely involved nutrients—calcium and vitamin D—which must work together for good health in young and old alike. Most mothers and pediatricians realize that babies need milk for its calcium and vitamin D to help their little bodies use the calcium. Babies are taken out into the sunlight so that additional vitamin D can be absorbed through their skins. But few people pay enough attention to the fact that old people need exactly the same kind of treatment. You never outgrow your need for calcium and vitamin D.

CHAPTER 5

The Ancient Disease of Rickets Is Still with Us

FOR MANY, MANY YEARS research scientists, nutrition experts and physicians have all been well aware of the importance of vitamin D and calcium for the prevention of rickets. So it is startling to encounter, as we recently did, an article in *The Lancet* for May 28, 1976, describing a number of cases of florid (that is serious or fully developed) rickets in a group of young children in Glasgow, Scotland.

Six investigators from the Royal Hospital for Sick Children and University Departments of Biochemistry and Child Health in Glasgow found that 12½ per cent of all the children they examined suffered from rickets, slight to very serious cases. Rickets is a disease of bone development. It causes knock-knees, twisted spines, deformed legs, arms and wrists. Because it can also deform pelvises, it can create untold hardships for women in childbirth if they suffered from rickets when they were babies.

Say the Scots authors, "It is unacceptable that fetal rickets should occur in Glasgow in the 1970's." Why and how did it happen? Practically all the children discovered to have rickets were African or Asian immigrants living in more or less segregated communities in Scotland, eating mostly their traditional food and, many of them, unable to converse in English. They shopped at their own stores, went to movies and church services conducted in their own languages and had little to do with the Scots teachers or nutrition experts or child clinics conducted in English.

But how does it happen that just not knowing the language can produce a serious, deforming disease whose effects will probably persist for a lifetime? The diet of the Asian and African immigrants consists mostly of chapattis (pancakes) made from wholegrain flour. With these they eat clarified butter, or *ghee.* People in hot countries who have no refrigeration preserve butter by melting it and removing all the cloudy part of the melted butter. What is left is pure fat which keeps well without becoming rancid, for some time.

The Glasgow doctors speculate that perhaps this cooking process destroyed what little vitamin D the butter contained. **There is no vitamin D in wheat flour. So these immigrant children were getting almost no vitamin D. The official Recommended Dietary Allowance for vitamin D in the USA is 400 units daily.**

Thirty-two per cent of the Asian children, 24 per cent of the Chinese, seven per cent of the African children were getting less than 50 units of vitamin D per day. Forty-six of the Asian mothers had stopped giving their children vitamin supplements by the age of one year. Most of the Asian and Chinese children generally retain their traditional eating habits into adulthood.

"The finding of rickets of all types in 12.5 per cent of all Asian children confirms a serious problem which warrants further action," say the authors, ". . . The solution probably lies in improved education of Asian immigrant mothers to teach them English and then the elements of child care in the British climate."

And therein probably lies the answer to this mystery of so many cases of a vitamin deficiency disease in immigrants from the South who come to a northern city in a northern country like Scotland. The usual weather in Scotland is cool, cloudy and rainy. The immigrants came from countries where the sun shines all day most days, the year around. The sun is high in the sky and many, many hours of full sunlight fall on the nearly naked skins of the African and Asian children playing outside for much of the day.

In Scotland weather necessitates wearing heavy clothing which shuts sunlight away from the skin. Cloudy skies and rain mean that very few hours of sunlight are available compared to the sunlight these children enjoyed at home in their native lands. In addition, these children are living in Glasgow where air pollution from furnaces and industries undoubtedly shuts off what little sunshine there is, especially during fall and winter.

Sunshine on the bare skin creates vitamin D. This is the best source of vitamin D. It is believed that the dark skin of Africans was developed over many generations to protect them from absorbing too much light, hence too much vitamin D from the blazing African sun. When they migrate north, the dark skin goes along with them. Here, where sunlight is uncertain and weak the dark skin continues to shut out what little sunshine there is. So people with dark skins from southern lands are always well advised to get plenty of vitamin D in food and food supplements when they go to live in the north.

No one told this to the African and Asian mothers in Glasgow, since no one could communicate with them. There was no way for these devoted but illiterate mothers to know that the cold, rainy, cloudy days were depriving their children of a life-saving bounty of vitamin D from sunlight on their skin. So they had no way of knowing that the vitamin D must be provided in some other way or the children would get rickets.

The same is true of native Scots children living in Glasgow whose mothers do not realize the value of getting vitamin D in food and in supplements. Several Scots children with rickets were reported in this same study. In Glasgow a short cartoon film has now been made in Hindi, Urdu and Panjabi languages emphasizing the importance of giving children suitable foods and supplements of vitamin D. It is shown at meeting places and to older children at school. Leaflets and posters in Asian and African languages are also posted in these communities. The doctors suggest, in addition, that supplements of 300 units of vitamin D should be made available to immigrant children and should be given daily between the ages of 3 months and 16 years.

They also suggest that the flour used by the immigrants to make their chapattis should be fortified with vitamin D. One of the reasons for this is that the immigrants' diet, which consists almost entirely of wholewheat flour mixed with water and made into pancakes or chapattis contains phytate which has a tendency to unite with calcium and other minerals and cause them to be excreted unused. **This is one reason why we do not recommend unyeasted bread. When yeast is used to raise wholegrain bread it changes the form of the phytate so that it does not cause this loss of minerals.**

But the African and Asian women have no knowledge of making bread with yeast and they are accustomed to eating unyeasted chapattis so they continue to eat them in Scotland. This, combined with the almost total lack of vitamin D in sunshine and diets creates the calcium deficiency that results in rickets. **Vitamin D is very scarce in foods.** We were apparently meant by nature to get most of our vitamin D from sunshine. But how can this be done in a far northern country by people whose skins are genetically developed to exclude sunlight?

Most of us do not have problems of dealing with immigrants and trying to solve their nutritional problems by understanding all these various aspects of vitamin D nutrition which change drastically when the individuals move to another part of the world. However, we do have our problems with vitamin D no matter where we live in the United States.

Rickets afflicted large numbers of children in early America, because no one then understood the cause of the disease. Egg yolk and the bones of fish like salmon, sardines, tuna, mackerel and so on, as well as fish livers, are the only reliable sources of vitamin D in food. There is a bit of vitamin D in milk, butter, cream, cheese, liver and meat, and a very small amount in salad oils.

Most dairy milk these days is fortified with vitamin D. The container says on the label that one quart of milk contains 400 International Units of vitamin D. This was recommended to prevent the epidemics of rickets that prevailed in earlier days. But what about people who don't or can't drink milk? Many people these days have been told by their doctors to shun both milk and eggs because of their supposedly harmful cholesterol content. Where will such people get their vitamin D? **It is now well known that adults as well as children need vitamin D to keep their bones healthy.**

Where can vegetarians who eat no animal products of any kind get their vitamin D especially if they spend most of their time indoors or in northern climates? **A diet which contains no foods of animal origin is quite likely to be short on calcium as well, so this is a double hazard in the badly planned vegetarian diet.** Vegetarian children who eat no animal products should get their calcium and vitamin D from supplements.

Its vitamin D and calcium content is one very good reason to use plenty of milk, no matter what your age happens to be. Milk is not just a food for infants, any more than eggs are food just for embryo chicks. See that your family gets plenty of milk and other calcium-rich foods such as all kinds of cheese and yogurt.

And see that your family gets plenty of vitamin D. If you live in the south it's much easier, for you usually have plenty of sunshine all winter. But your skin manufactures vitamin D only when it is bare to the sunshine. So don't bundle your children up unless it's really cold. Let them play outside all summer with as little as possible between them and the sky. They need not, and should not, stay in the sun long enough to be sunburned. You get plenty of vitamin D just from being in the open air on a sunny day, even if you spend all your time in dappled shade. Sunburn is unwise.

And take a vitamin supplement if you think you are not getting enough vitamin D. The official Recommended Dietary Allowance is 400 I.U. daily. Vitamin D is a fat soluble, so it accumulates in the body. You need not take it every day. And don't take too much; 4,000 units or more a day can produce symptoms of overdosage: lack of appetite, thirst, diarrhea, nausea, joint pains and so on. But be sure you and your children get enough, and get enough calcium to guarantee strong, well formed bones and teeth.

CHAPTER 6

Osteoporosis, the Soft-Bone Disease

SEVERAL YEARS AGO, a university specialist told a New York audience that osteoporosis is bound to become a growing problem, since the 50-plus age group is the fastest growing minority in the United States. Osteoporosis is the softening of bones that may occur as we grow older.

It affects the entire skeleton. The back, the legs and feet are affected. As the spine collapses with time (because the small bones are too weakened to support it), the ribs may fall forward onto the rim of the pelvic bones, thus crowding all the digestive organs and creating much distress. Osteoporosis is involved with the actual loss of bone.

This disorder is largely responsible for the more than one million bone fractures each year in women over 40 years of age. "Seven hundred thousand of these occur in women with osteoporosis," said Dr. Louis V. Avioli, Professor of Medicine at Washington University in Missouri. "Seven hundred thousand of these fractures occur in women with osteoporosis, and 25 to 30 per cent of all post-menopausal women have the ailment," he continued.

Osteoporosis is more common among women than men. The average man loses five to six per cent of his bone mass every 10 years after the age of 35. At 65, bone loss slows to two to three per cent every 10 years. In women, the rate of loss is doubled. And 25 per cent of American women of 40 are losing bone at a faster rate than other women their age.

A number of everyday things, including diet, are probable causes of bone loss, said Dr. Avioli, but not even the experts know which is the most important or why. Women have decreased levels of the sex hormone estrogen after menopause. This seems to be one reason for bone loss. (We would point out that recent research seems to show that taking estrogen in the form of a hormone pill may produce a cancer, so it does not seem wise to try to replace the lost estrogen).

Deficiency in calcium is undoubtedly another reason. This can be caused by not getting enough of the mineral at meals or not absorbing it completely. Many surveys have shown that a large percentage of American women, especially those over 40, are not getting even the minimum recommended amount daily. It is almost impossible to get enough calcium without using lots of dairy products, especially as we grow older and tend to absorb less of the minerals in the food we eat. A glass of milk or a big chunk of cheese at every meal, along with calcium tablets will go a long way to make up for such deficiency.

Vitamin D is essential for the body to absorb calcium. We get vitamin D chiefly from the action of sunlight on bare skin. This suggests that women in northern climates get very little vitamin D in winter, fall and early spring. However, many dairy foods these days are enriched with some vitamin D. Most milk contains 400 International Units per quart. It is also possible to take a natural vitamin D supplement during the winter. This is fat soluble, so it is not necessary to take it every day. It is stored in the body until it is used up, so once a week is enough. It is toxic in very large amounts, taken over long periods of time. Take the natural kind of vitamin D from fish liver oils.

It is also essential to have some fat in the digestive tract at the time you eat calcium-rich food, if the food is to be well absorbed. On the other hand, too much fat at the same meal will cause the calcium to form "soaps" with the fat and be excreted without being absorbed. What to do? Don't depend solely on nonfat milk for your calcium. And don't depend on great blobs of cream or whipped cream for your calcium, as these may have too much fat for good calcium absorption.

Dr. Avioli also mentions the importance of "muscle mass" which, he says, peaks at around 20 years of age. "The day-to-day pull on muscle tendons stimulates bone formation," he said. Inactivity also causes one to lose bone structure. The minerals just ease out of bones when we are confined to bed for long periods of time or when we spend most of the day sitting. This suggests that daily exercise is essential for maintaining healthy bones and avoiding osteoporosis.

It seems also that there is a definite relationship between loss of bone and a diet in which there is too much protein in relation to calcium. **If one is making use of dairy products which contain both protein and calcium, there seems to be no danger of this.** But people who, for some reason, shun dairy products may eat lots of meat, which contains no calcium. Thus they throw the relationship out of balance. It is the acid-ash of the high-meat diet which destroys the calcium level, according to Dr. Avioli. We would point out that a diet high in sugar also creates an acid-ash. So avoid the use of sugar if you would save your bone structure into a healthy old age.

Dr. Avioli tells us that about 6.3 million people in this country are at present suffering from acute problems relating to weakened bones in the spine. And eight million Americans have chronic problems related to the spine, compared to only six million in 1963. Our situation in regard to bone health is obviously getting worse year by year.

Then, too, recent surveys show, said the Missouri specialist, that a minimum of 10 per cent of all women over 50 suffer from bone loss severe enough to cause hip, vertebrae or long-bone fractures.

"Surveys in homes for the aged and of ambulatory patients 50 to 95 years old have disclosed symptomatic backpain osteoporosis in 15 to 50 per cent of these populations respectively."

It's nothing to ignore, hoping that it won't happen to you as you grow older although you may have seen it happen to older relatives and friends. Dr. Avioli suggests the following means of prevention. Plenty of activity and exercise. **Don't allow yourself to become housebound or chair-bound or bedridden. Get out, move around, get interested in some hobby involving vigorous exercise.**

"Almost 100 per cent of everyone over 45 eventually has periodontal (gum and bone) disease and by the age of 65 almost half of all Americans have lost all their teeth . . . the serious nature of bone loss in the jaw is as yet not fully realized. It results in decreased tooth support, root fenestration (root lacks bone support) and actual loss of teeth," reports Anthony A. Albanese, New York specialist, in *Food and Nutrition News,* October-November, 1975.

Dr. Leo Lutwak is convinced that you can predict a case of osteoporosis by examining an individual's mouth and studying the condition of his jawbones. In *Nutrition News* for February, 1974, Dr. Lutwak, who is professor of medicine at the University of California in Los Angeles, says that **examination of the jawbone by a dentist will reveal degeneration of the bone which is responsible for what is defined as periodontal disease—that is, any disease which affects any of the tissues surrounding the teeth.**

When calcium is lost daily from the body, due to disorders in the glands which regulate calcium and/or due to lack of calcium in meals, the jawbone loses calcium, the teeth begin to move about in their sockets, chewing causes irritation and damage to the gums. The gums bleed, become inflamed and are readily infected with plaque—those thick scales which dentists believe are very destructive of teeth and gums alike.

If dentists recognize this disease of mouth tissues for what it is—a disease of calcium deficiency—they can and should advise their patients to get more calcium. The gum condition develops before other manifestations of bone softening appear in hips, spine and legs. Alerted by the dentist, the person whose calcium intake has been too low can, Dr. Lutwak believes, increase his intake of calcium and, with it, his chances of escaping a full-blown case of osteoporosis which brings with it pain and disfigurement and, usually, broken bones.

It's a very attractive theory, especially for those of us who are mostly concerned with doing everything we can to prevent diseases, rather than trying to patch them up after they happen.

How is it possible that so many aging Americans could be so short on calcium that they face a painful old age due to osteoporosis? Easy, says Dr. Lutwak. **During the past 15 years or so, Americans have been eating more meat and less dairy products.** The meat is a good source of phosphorus, but contains no calcium. Dairy products are the best source of calcium. So the ratio between these two minerals has been changing over the past years.

In 1960 this ratio was about 1:2.8—that is, the average American ate 1 part of calcium to every 2.8 parts of phosphorus and this was a good ratio. By now, however, this ratio is more like 1:4—too much phosphorus for the amount of calcium. We need to increase the calcium part of this ratio—that is, eat more foods that are rich in calcium or get more calcium in food supplements.

Dr. Lutwak explains how our bodies use calcium. It is carefully metered out by a complicated mechanism involving the parathyroid glands, the pituitary gland, the adrenals and the thyroid. Most of it is deposited in our bones. Only about one per cent remains in extracellular fluid. The official recommendation for daily calcium intake for adults is 800 milligrams. **"Many dietary surveys show," says Dr. Lutwak, "that the vast majority of American adults—and particularly women homemakers—consume only about 400 milligrams daily. Of this they may absorb only about 10 to 15 milligrams.** So they may have a negative balance of 90 milligrams of calcium daily. Over the course of 30 years, this mounts up to 980 grams of calcium lost. In other words, by age 50, only one-third of all the calcium originally in the bones is left! So no wonder bones begin to soften and the painful symptoms of osteoporosis begin to appear.

Hospital records show that just about everybody who comes in with osteoporosis has also a case of periodontal disease. A group of 90 such patients were selected for an experiment to see if the further progress of this disease could be stopped just by getting more calcium. The patients were divided into two groups, one of which got a calcium supplement daily, the other getting a tablet which looked the same but contained nothing.

The group which got the "nothing" pill showed no change in the condition of their jawbones over a year. The group which got the calcium supplement showed about 12½ per cent increase in bone density—that is, stronger, thicker bones, much less likely to succumb to osteoporosis.

Dr. Lutwak wants to encourage dentists to examine the condition of their patients' jawbones very carefully and recommend calcium supplements to those who are beginning to show wear and tear on the bone. People visit their dentists regularly, he says, and dentists should perform this service. But, of course, the people themselves must implement his suggestions. **They must begin to get more calcium at mealtime and/or take calcium supplements every day to prevent the further spread of the degenerative disease osteoporosis.**

A doctor writing in *Internal Medicine News* for March 1, 1974 says he believes that one reason for bone fractures in old folks (aside from lack of calcium) is that they do not spend enough time on their feet to maintain bone strength. Our spines, legs and feet were designed to bear weight. When, day after day, year after year, we do not use them for that purpose, they degenerate, as do other parts of us that do not get enough use. So let's get out and walk in the open air to get back the use of our legs.

Pediatrics for July, 1974 reports on a serious bone disorder, *Osteogenesis imperfecta*, in which bones fracture following only minor bumps or bruises. Researchers gave these patients 1,000 to 2,000 milligrams of vitamin C daily and found they had less tendency to break bones. Why not go a step farther, gentlemen, and suggest that we all take vitamin C to prevent a tendency toward bone-breaking, especially if the tendency runs in the family? And what's wrong with giving far more than 2,000 milligrams just to see if these fracture-prone patients can manage to get by with no more fractures? Maybe much larger doses are required.

The Lancet for June 28, 1975, reported on a survey of the vitamin D status of 56 women patients over 65. They all had very low blood levels of vitamin D. Since this vitamin is essential for the absorption of calcium, it seems likely that the condition of these patients' bones was not good and was likely to become much worse.

If you are taking calcium supplements to prevent or treat osteoporosis, it might be wise to take the calcium at night, just before retiring, rather than during the day. This is the theme of an article in the *British Medical Journal* for June 2, 1973. Four physicians from the London Hospital in England tell us that it seems most calcium is lost from bones during the night, in cases of osteoporosis. Apparently the calcium that you may take with meals is absorbed within three hours after you take it. So by midnight any effect of the evening meal in raising calcium levels of the blood would have ceased.

In women past the menopause, it seems calcium is needed to suppress the action of a certain hormone

which tends to cause bones to soften. If most of the blood calcium is gone before retiring, therefore, the hormone could perform its debilitating work during sleep. Doctors have found that, before the menopause, women excrete much less calcium in the morning than women who are past menopause. This seems to indicate that something, overnight, causes the body to lose calcium (presumably from bones) in the woman who has passed menopause.

The London doctors did tests on a group of 49 women, giving some of them a supplement of calcium (800 milligrams in all), which they took just before going to bed. They were told to eat as usual during the day and, of course, to include foods with plenty of calcium in them, since lack of calcium in foods is associated with osteoporosis.

Then the doctors checked the amount of calcium excreted in urine during the night and immediately upon rising. They found that less was excreted when the calcium supplement was taken just before retiring. So they suggest that women past the menopause who are trying to prevent osteoporosis should take calcium supplements in one dose, just before going to bed, rather than extending it, in smaller doses, throughout the day.

Two French scientists, writing in *Presse Medicale*, November 9, 1963, state that one of the major causes of osteoporosis may be lack of vitamin D. A British Columbia physician, writing in the *Canadian Medical Association Journal*, January 23, 1965, says, "Vitamin D must be present to ensure adequate absorption of calcium." He suggests no more than 400 units of vitamin D daily.

"Osteoporosis in the aged is probably due to a combination of poor nutrition, loss of . . . sex hormones and inactivity or immobilization. In old people deficiencies of protein, vitamins and calcium may result from anorexia (lack of appetite), certain food habits, unbalanced diets, lack of teeth, food idiosyncrasies, serious and prolonged illnesses or economic conditions . . . there is a tendency toward low blood levels of thiamine (vitamin B1), ascorbic acid (vitamin C), carotene (vitamin A), and vitamins despite adequate intake, absorption and assimilation of these substances . . . (this) may be due to either an hepatic dysfunction (liver trouble) . . . or an increase in the physiologic need for vitamins beyond the requirements for adults and children," Dr. M. L. Riccitelli reported in the June, 1962 issue of the *Journal of the American Geriatrics Society.*

The scientific evidence for the importance of calcium in preventing and treating osteoporosis is voluminous. Here are some other statements from the wealth of material available.

"In individuals accustomed to low calcium intakes, osteoporosis is more common than in those who have had higher dietary intakes of this mineral . . . requirements of amounts of dietary calcium . . . increase with age in both men and women . . . Shorr and Carter were able to produce significant storage of calcium in some patients with osteoporosis by raising dietary calcium intake to 2 grams or more daily," *Osteoporosis, a Disorder of Mineral Nutrition*, by Leo Lutwak and G. Donald Whedon.

"It is now believed that bone formation is not defective in persons with osteoporosis but that there is a calcium deficit. Special attention should be given to . . . patients' diets, both by insuring plenty of foods with a natural calcium content and also by the addition of mineral calcium to them (that is, food supplements)," J. L. Newman in *Journal of the New Zealand Dietetic Association*, December, 1962.

"There are increasing data to suggest that in both sexes as good if not better results can be achieved with a high supplementary calcium intake as with steroids (sex hormones). The use of calcium is simpler, cheaper and less hazardous. A calcium supplement of 3 grams (3,000 milligrams) daily should be given. One gram of calcium is provided by four one-gram tablets of di-basic calcium phosphate. Eleven one-gram tablets of calcium gluconate or 13.6 gram tablets of calcium lactate will provide the same amount of calcium . . . ," Hamish W. MacIntosh, M.D., in the

Canadian Medical Association Journal, January 23, 1965.

"Many patients with these forms of osteoporosis absorb and retain calcium abnormally avidly when on a high calcium intake in the form of supplements and moreover may continue to retain it avidly for at least three and a half years. Symptoms of the disease are relieved and no further fractures take place," *The Lancet*, Vol. 1, 1961, page 1015.

"Calcium salts, in doses of two to three grams daily, have been shown to induce a positive calcium balance . . . serum (blood) vitamin D levels in osteoporotic patients are below normal . . . some bone biopsies have demonstrated changes which might be the result of a relative vitamin D deficiency; and . . . vitamin D administration has resulted in increased calcium retention," *The Journal of the American Medical Association*, May 18, 1964.

As we reported earlier, **there is considerable evidence that vitamin C also has something important to do with getting the calcium into the bone structure and replacing it when necessary**. We know that this vitamin is important for building bones in infants—why not in adults, too?

An article in the *British Medical Journal* for March 12, 1966 discusses a number of South African patients with osteoporosis who also had scurvy—the disease of vitamin C deficiency. Say the authors, "The association of osteoporosis with scurvy is well recognized and probably relates to the role of vitamin C in the formation of collagen and hence new bone."

Vitamin C is most abundantly present in fresh fruits and vegetables like citrus fruits, strawberries, melons, broccoli, green peppers, cabbage. It is also available in food supplements.

These, then, are rules for preventing osteoporosis: a diet high in protein, calcium, phosphorus, vitamin D and vitamin C. And plenty of exercise.

Osteoporosis is a very serious and painful disorder that affects many people past middle-age. It is also a very common disorder. How often have you heard that an older person—usually a woman—has fallen and broken her hip? Actually, the hip was not broken by the fall. The weakened bone in the hip, unable to withstand the stress of moving about, broke, and the fall followed. Doctors can sometimes mend these broken hips, following months or weeks of hospital attention. But isn't it more sensible to try and prevent such accidents from happening? This would also apply to legs, arms, and other bones which conceivably could break during a fall, simply because they also were too fragile to withstand the strain.

A useful leaflet entitled *Facts About Osteoporosis* is available from the Information Office, National Institute of Arthritis and Metabolic Diseases, Public Health Service, Bethesda, Maryland. A high-calcium diet is outlined here. It includes three 8-ounce glasses of milk, one ounce of cheese, three slices of bread, a serving of meat, one egg, two or three servings of vegetables, two servings of fruit and a dessert made with milk. The total calcium in such a day's diet is 1,350 milligrams, with 1,064 milligrams coming from the milk and cheese.

A new study has pinpointed smoking cigarettes as an additional hazard which may provoke osteoporosis or bone softening in older women. In *Archives of Internal Medicine*, Vol. 136, 1976, pages 298-304, Dr. H. W. Daniell describes a survey of women between the ages of 40 and 70 who were known to have certain kinds of osteoporosis. In an earlier article, Dr. Daniell had noted that middle-aged men and women with osteoporosis causing symptoms were mostly heavy smokers.

Comparing the health records of a large group of women smokers and non-smokers, Dr. Daniell found that the percentage of cigarette smokers was much higher (76 per cent) in those with osteoporosis than in a group of office patients without osteoporosis. Assays of bone density also showed that women with osteoporosis were more likely to be smokers.

The Journal of the American Medical Association, commenting on these findings, says that the Surgeon General's warning on packs of cigarettes may not now be sufficient. In addition to lung cancer, respiratory disease and other lesser hazards in smokers, we should now add the threat of additional harm from osteoporosis as we grow older. It's enough to make one douse the cigarette one has just lit and vow permanently that it's the last. Why not?

Dr. Anthony A. Albanese and his colleagues studied 12 elderly women who got calcium supplements and another similar group who took none. Over three years the women who took calcium showed increased density of bone (stronger, healthier bone), while the other group continued to show more and more bone destruction. **The scientists also found, surprisingly, that in the space of only one year, blood cholesterol levels went down in the women taking calcium.**

Then they studied 23 women from the age of 36 on up. Fourteen of these took three calcium tablets a day along with their regular meals, while the other women took three capsules which looked like the calcium tablets but contained nothing. In many of the women taking calcium the condition of their bones improved so greatly that they approximated that of men of similar age. And women are known to be more prone to osteoporosis than men are.

These doctors concluded their studies had shown beyond a reasonable doubt that we need a daily minimum of one gram of calcium (1,000 milligrams)—and perhaps more—just to hold our own and maintain normal bones as we age.

Dr. Herta Spencer of an Illinois Veterans Hospital has also been studying calcium in relation to bone health. **She believes that 1,200 milligrams of calcium may be necessary to maintain good bone health.** In her experiments, Dr. Spencer involved people without osteoporosis and people with it. Dr. Spencer found that only some could maintain fairly healthy bones with 800 milligrams of calcium a day. All of them did well on 1,200 milligrams daily. She supplemented their daily diets containing 200 milligrams of calcium a day with milk or calcium gluconate supplements.

She concluded that, "Because it is not yet possible to predict the development of osteoporosis and because of the difficulty in diagnosing and treating this condition, it is desirable for adults to consume 1,200 milligrams of calcium daily rather than the presently recommended dietary allowance of 800 a day."

She also found that the antacid drugs taken by millions of Americans prevent calcium from being absorbed, so that bone loss is almost certain even in people who go out of their way to get enough in their diets. Ulcer patients, for example, taking one or another of the antacid drugs in amounts much smaller than most ulcer patients take, suffered significant loss of both calcium and phosphorus. More and more of these minerals must be provided or bones will surely suffer.

CHAPTER 7

Immobilization Causes a Loss of Minerals

IF YOU HAVE ever visited a rest home or nursing home, you have probably been appalled at the number of older people who appear to be completely immobilized. Although conscious and not suffering from any acute disorder, they lie in bed constantly or they sit in wheel chairs from which they are apparently never moved, except to return to bed.

An important study of this situation was reported in the August, 1975 issue of the *Journal of the American Geriatrics Society*. Michael B. Miller, M.D., FACP, discusses the eventual effects of such immobility on nursing home patients. They are horrible to contemplate. And they can be reversed by special therapy. There is no need for prolonged inactivity of most patients. And their immobilization makes more work for the nursing staff and doctors, as well as more expense for everyone concerned. Dr. Miller calls these sad effects of immobilization "iatrogenic" (which means caused by doctors) or "nurisgenic" (caused by nurses).

He tells us first of the great importance of two things in

preventing harm to bones—first the mineral calcium in ample amounts every day. **There is no indication that our needs for calcium decrease as we age. Indeed they seem to increase.** The second most important element in the quite serious disorder which affects older people who are immobilized is the lack of what doctors call "skeletal stress." It seems that bones, to be healthy, must be in constant use. This is especially true of those which bear weight—the pelvis, hip bones and leg bones. They were designed by nature to bear weight. As soon as this weight is removed and the individual lies motionless in bed or sits in a wheel chair, these bones begin to disintegrate. If there is an accompanying lack of calcium, osteoporosis is almost certain. Osteoporosis is the softening of bones, as minerals are withdrawn and the bones become unable to bear weight.

Calcium, phosphorus, potassium and other minerals are lost in urine and feces when one is immobile. This makes the situation worse. It is well known that circulatory troubles also increase during bed rest, for the simple reason that the valves in the legs do not function properly unless one is up and about. Walking purposefully and briskly is the best way to keep these valves pumping blood along the blood vessels in the leg so that it does not accumulate, become sluggish and clot. The additional loss of so many minerals from bones compounds the health problems.

But, as Dr. Miller tells us, prolonged immobilization affects the personality and the social outlook of the bedridden patient. He gives us case reports of six elderly women who were put to bed in a nursing home because of broken hips, heart disorders, infections, amputations and so on. After four weeks of immobilization, they had apparently decided they were dying and they began to behave appropriately. They stopped eating and drinking fluids. They stopped talking. They refused to communicate in any way with nurses or doctors.

If nurses tried to feed them, they spit out the food. Several of them became incontinent of urine and feces. If they were placed in wheelchairs they deliberately caused accidents. When nurses attempted to get them to stand, they went through a series of bizarre movements all calculated to prevent themselves from standing. If they were helped up, their knees collapsed beneath them and they fell to the floor.

Throughout all this there was no measurable evidence of any damage to legs or nerves. True, muscles had wasted and bones were deteriorating because of the prolonged bedrest, but the personality changes, too, appeared to have no other basis except the fact that they had been confined to bed for so long.

Says Dr. Miller, "the syndrome is reversible." These old folks can be gotten to their feet. They can be rehabilitated. They can learn to feed themselves once again. They will once again begin to communicate with those around them and become part of the social life of the establishment. After their rehabilitation they will admit that they thought they were dying and resigned themselves to it. The only reason for this conviction was the total effect upon them of prolonged immobilization.

Dr. Miller says, "In the absence of continuing and direct medical involvement in the progressive rehabilitation of the severely disabled aged patient in whom total disability is exacerbated by the onset of acute illness, the nursing staff in a long-term care facility must assume responsibility for nursing rehabilitation even when such maneuvers require the removal of restraints ordered by physicians. Responsiblity for the safety of the patient thus accrues to the nursing staff.... latrogenic (doctor caused) factors in producing the patient's disability have long been recognized. Nurisgenic (nurse-caused) factors are now coming to the fore."

In other words, doctors in general are not aware of the damage being done to their long-term patients who are immobilized. Or if they are, they seem not to know what to do about it. So it is the job of the nurses who are with these patients all day to get them out of bed and out of wheelchairs for their own salvation. Of course, as they become, once again, able to care for themselves, to bathe, clothe and feed themselves, the work of the nurses is greatly decreased.

The reason for the original immobilization, which brings all these terrible consequences is, usually, breaking a bone, having a heart attack or some other circulatory disorder or related disease which disables the older person. **So the sensible person will do everything possible to avoid such health disasters. For, at any age, prolonged immobilization will bring severe side effects.**

It is believed that the broken bones which accompany falls in older people are caused not by the fall itself. Rather the bone just disintegrates and becomes so fragile that it cannot support weight. So it collapses and the individual falls. Plenty of minerals in the diet—chiefly calcium—are the only ways to prevent these emergencies.

We have discussed food sources of calcium in another chapter. Then there are calcium supplements. All are recommended. Phosphorus is another mineral which balances calcium in the body and generally the more phosphorus we get the more calcium we need. So if your diet is high in phosphorus, it might be a good plan to take a food supplement of calcium alone. If you get less phosphorus, then bone meal, or some other all-around mineral supplement would be suitable. These contain phosphorus in addition to calcium.

It helps to have some fat in your digestive tract at the same time as the calcium, so you probably absorb more calcium from whole milk than from skim milk. On the other hand, too much fat causes you to lose calcium. As always, be moderate.

Vitamin D is essential for the absorption of calcium. You get vitamin D from sunlight in summer and spring, if you are outside part of the day. In winter, especially in northern states, it's a good idea to take some vitamin D—every week or so—since it is fat-soluble and your body retains it well.

CHAPTER 8

Milk

ONCE IN A WHILE, you encounter someone who hates milk, hates it with a passion, which he has fortified with much research, he tells you, proving that milk is food only for babies and that anyone over the age of two has no business drinking milk. If Nature had meant us to drink milk, she would have provided it from a tree or some other source, he may tell you. Any animal's milk is food only for that animal's offspring and man is doing himself grave harm by drinking it. Since we are editors of several health-oriented magazines, we often receive such "hate mail."

No doubt many people down through the ages have felt this way about milk. Certainly there have been many cultures where milk other than human mother's milk was never used. But mostly, it seems, this was simply because there was, in such cultures, no way to raise or feed milch animals. **In any case, we know that milk-drinking goes very far back in human history and that many nations have survived and lived in a state of excellent health drinking milk, sometimes along with an excellent diet, sometimes using milk as almost their only food.**

In a fine book, *Food in Antiquity*, by Don and Patricia Brothwell, we are told that in a frieze at Ur—an ancient Sumerian city on the Euphrates river—which must have been around 2900 B.C., there are pictures of human beings milking cattle. So we know that at least 5,000 years ago people were drinking milk. The early Greeks used goat and sheep milk. Pliny, the Roman naturalist, thought that camel milk was the sweetest. In northern communities, reindeer milk has been used from very early times. The elk is also domesticated and milked in some parts of the world. Yaks and asses, buffalo and camels have given milk to many nations in past history.

Since there was no way to refrigerate milk in early days, butter, sour milk and cheese were probably discovered rather soon after early humans domesticated animals. Any milk left to stand would separate into cream and milk. The milk would sour. If the cream were transported from place to place, it would be churned in the process to turn it into butter.

In some early Egyptian tombs, the fatty substance that has been found is believed to be butter—thousands of years old. The Bible mentions butter. Some later experts believe the word used may mean "curds"—the food that results when milk is coagulated. In Greece and Rome in classical times, butter was considered food for barbarians. The Greek and Roman gourmets used olive oil as their chief source of fat.

But the Romans also used soured milk much like our yogurt. They called it *oxygala* and it must have been fairly solid, as most of the whey or liquid part was drawn off. They made another soured form of milk called *melcaby* by pouring fresh milk into jars containing boiling vinegar and keeping it overnight in a warm place. Their gourmet chefs prepared it with salt, pepper, oil and coriander.

From Neolithic times, when human beings settled down into communities and began to plant and harvest crops, they have made cheese. The Egyptians made it. The early Greeks and Cretans made it. Homer talked of a pottage made of barley-meal, honey, wine and grated goat milk cheese. The Romans smoked cheeses and imported foreign varieties. It was also an important ingredient in their cakes and breads. The Greek peasant in classical times had a basic diet of barley pastes and gruel, barley bread, olives, figs and goat milk cheese, washed down with goat milk.

Different kinds of cheeses and curds were eaten in ancient Assyria and Babylon. They were served in a variety of shapes in moulds which have been discovered by archaeologists. This highly sophisticated civilization in Mesopotamia (between the Tigris and Euphrates rivers) began about 2900 B.C. and lasted until the invasion of Alexander the Great in 330 B.C.—a vast period of time of which we have a great deal of information due to the record-keeping proclivities of these people.

Reay Tannahill says in *Food and History* that the nomad herdsmen of the first thousand years A.D. considered horses their wealth, for the mare's milk "may well have been a decisive factor in the nomads' exuberant good health." **They ate plenty of meat and milk to obtain protein and the B vitamins, as well as calcium.** Since fresh fruits and vegetables were almost unobtainable in this kind of life, they would have been short on vitamin C except that mare's milk has twice as much vitamin C as human milk and four times more than cow's milk.

Koumiss is the soured milk product of many nomadic peoples of the Near East. It is much like our yogurt and it was apparently one of the earliest foods of these ancient peoples. An early writer tells us that koumiss was made by pouring fresh milk into a great bag which was then beaten with a piece of wood until the milk became frothy and soured. (The milk had been fermented before this). The churning went on for three or four days. The Kazak people today still make their koumiss this way in Russian Turkestan.

The dowry of an early Persian princess included 100 milch camels and 100 camels "for burdens." Betty Wason in her book, *Cooks, Gluttony and Gourmets*, tells us that in the "Shah Nameh," which is a collection of early legendary tales, one early Persian king was said to have 1,000 animals which gave milk—goats, sheep, camels, but no cows. These early people never drank the milk fresh and sweet. They preferred buttermilk, with the cream skimmed off. Or they might eat the curds of the buttermilk—the yogurt—as it was, beaten with fresh spring water as a foamy drink somewhat like an ice cream soda without the carbonation.

The solid part of milk—the yogurt—was used in many early recipes. Mixed with minced onion, black pepper and herbs it was a sauce for shish-kebab, meat grilled on a spit. Adding minced cucumbers and raisins, the ancients made a creamy cold soup. By adding honey or fruit marmalade, they had a dessert. Or it might be served as a salad dressing over crisp cucumbers and radishes, both well known and widely used vegetables. All of these recipes are just as tasty today using modern yogurt as we make it in our modern kitchens.

The earliest mention of buttermilk in India occurs in a Sanskrit manuscript "Tales of Ten Princes." In one story a prince seeks a bride by asking all the girls in the neighborhood to cook him a meal with the bag of rice he had with him. The young lady who won his heart served him buttermilk along with a soup and the flavored rice dusted with cinnamon. Yogurt was an ancient food in India. They eat quantities of it in that country today. *Ghee*, their shortening which is used in almost everything they cook, is made by melting butter, then removing every bit of solid fat until only the liquid remains. This will keep for some time without refrigeration.

Galen, the early Greek physician, treated some illnesses with milk and insisted that when asses' milk is used, the animal must be brought into the sickroom. Considering the dangers of contamination, this seems an excellent idea. And some later physicians thought that human babies should nurse directly from the milk animal to prevent any contamination. On the other hand, there were those who were convinced that the milk a young creature drinks will greatly influence the way he turns out. There were apocrophal stories of pigs raised by mother dogs, which behaved exactly like dogs when they were grown. Might not a child raised on cow's milk turn into something like a cow later on? Of course, there was no way in those early days to analyze milk or to discover much of anything about its nutrients or their value. It is well to keep in mind that, in early times, the child whose mother died or could not nurse it died, unless some form of animal milk was available.

As human beings congregated more and more in villages and cities, the problems of providing milk to these urban people were immense. Molly Harrison, in *The Kitchen in History*, tells us that there was no control over the sale of milk, even by the 18th century in England. "When it came from farm cows, it was sometimes fresh, but often sour. Town cows were kept in such appallingly dirty conditions that their milk must have been the most dangerous of foods. Milkmaids carrying open cans of milk on their heads went from door to door, the open cans protected their heads from the slops and rubbish thrown from upstairs windows, but no housewife or cook can have bought milk with any confidence."

Methods of canning food were invented by a Frenchman named Nicholas Appert, who finally produced food in sterilized sealed cans, which would keep for years without spoiling. He won the prize which Napoleon had offered for such an invention. Later Louis Pasteur, who first told the world about bacteria and their important role in food as well as in illness, discovered that, if you heat milk to a low temperature which will destroy most of the bacteria, the milk will keep much longer than raw milk. This is the reason why most of the milk we buy today is pasteurized—the name, of course, comes from its discoverer. Most of our milk today is also homogenized, which means that the cream has been mixed throughout the milk, rather than in olden times when the cream always settled at the top of the container.

Canned milk was invented in 1835, but Gail Borden (that famous name in the dairy world and an American) improved on the method by introducing condensed milk which had been sweetened. This became popular with soldiers during the American Civil War. Until scientists understood what was involved in sterilization, the sugar was added because it helps to prevent the growth of some bacteria. And the early condensed milk—perfectly safe from contamination—was a much better food than any milk available to most people who did not have their own cows.

But the cheaper brands of condensed milk were made from skimmed milk so they contained no vitamin A or D. Children fed exclusively on such milks were bound to suffer deficiencies in these two essentials. Probably this was one reason for increase in rickets in poorer areas.

If you happen to be one of those people who just can't drink milk—that is, it gives you indigestion—try yogurt or cheese, either of which will give you all the nutrients of milk in a form which will cause no difficulty with digestion. The reason some people—many Orientals, for example—cannot drink milk is that they never had any after they were weaned. The enzyme *lactase* is present in all infants' digestive tracts to digest lactose, the sugar in milk. In people who do not drink milk after weaning, this enzyme disappears. In those who continue to drink milk, this enzyme remains active.

But in making yogurt, koumiss, buttermilk, cheese or any other soured milk products, the character of lactose is changed and the missing enzyme is not needed. So if you have had trouble in the past digesting milk, you should have no trouble with any of these soured milk products.

For those who cannot drink milk, most health food stores now have several products which provide the lactase enzyme. One such product, derived from baker's yeast, is in powder form which you add to a quart of milk according to directions. Another product, which consists of lactase and rennin, aids in the digestion of milk, yogurt, kefir, buttermilk, cottage cheese, etc. It is in tablet form.

As we indicated, there are millions of people around the world—Orientals, Africans, etc.—who simply cannot drink milk. Relief agencies in the underdeveloped countries don't help matters by sending the people powdered milk. They become nauseated and vomit, have diarrhea and cramping.

In the May 5, 1975 issue of the *Journal of the American*

Medical Association, there appears an article on this subject by Dr. Laurence M. Hursh, Director of Health Services of the University of Illinois. Dr. Hursh is objecting to the milk industry's claim that "Milk has something for everybody." The content of his article, however, shows a curious agreement with this claim, in terms of the actual nutrients milk offers.

"Milk contains a more versatile array of nutrients than most single foods," he says. "Thus milk is one of the best foods to balance the nutritional adequacy of a meal. As a food for infants, milk has no equal. As a source of calcium, of course, milk not only balances the diet, it is the chief source of calcium in the American food supply: 76 per cent of the calcium available to Americans in their food comes from milk and its products. If you doubt this, just try putting a day's meals together without milk or its products and still have the recommended daily allowance for calcium for any age level."

He goes on to say that **milk is the nearly perfect food because it contains essential nutrients in exactly the right balance**. The calcium it contains is easily available to the body because milk also contains just the right amount of phosphorus plus the right amount of vitamin D to help us to absorb both calcium and phosphorus. The protein of milk is so excellent that casein, the chief protein, is used as a standard reference point on which to evaluate the usefulness of proteins of other foods. They are compared to see how complete they are. Vitamins and minerals and healthful fats are also present in milk.

Since there are those people who lack the enzyme lactase, scientists got busy to test just why such people have difficulty drinking milk. As we indicated earlier, they found that milk sugar (lactose) needs lactase for good digestion. So they tested these people who had difficulties with milk in a laboratory, giving them immense amounts of lactose (100 grams) in water, on an empty stomach. They got sick. This is the way laboratory tests must be done, it seems. But it would take half a gallon of milk to provide 100 grams of lactose, so this test had almost nothing to do with the way milk is drunk in real life.

Dr. Michael Latham, Professor of International Nutrition at Cornell University, Ithaca, New York, has demonstrated in recent research, says the *JAMA*, that even people who cannot take large doses of lactose "can consume nutritionally useful amounts of milk" without discomfort. They can easily digest one glass of milk (with about 12 grams of lactose), especially if it is taken along with other food.

The problems that develop in places like the Philippines are elaborated on in a letter to the *American Journal of Clinical Nutrition*, April, 1975. Carol B. Suter, R.D., foods and nutrition specialist of Texas A. and M. University, College Station, explains how milk available under the Care program was being handled in the Philippines. The children had been eating a mixture of powdered milk, sugar, margarine and cocoa in candy form and no one had any trouble with the lactose in the milk. Apparently there was little milk used and its effects were diluted by the other ingredients in the "candy." The powdered milk was then given with one part of milk to four parts of water. Diarrhea resulted. The formula was changed to one part of milk to 12 parts of water and everything was fine. Gradually, Ms. Suter increased the amount of powdered milk until it was, finally, one part to five parts of water, and from then on there was no trouble with diarrhea.

In answer to Ms. Suter's letter, three Cornell University scientists reply that they, too, have found cases of intolerance to milk when there was just too much powdered milk in the formula. They also found people who could easily drink milk if it was taken along with other foods—a banana, for instance.

If you or your children cannot drink milk, we recommend that you dilute the milk with water and take it always with other foods, never alone. This dilutes the amount of lactose that enters the stomach at one time and, gradually, it seems likely that any individual may build up a supply of the enzyme lactase, and there should be no further difficulty.

A much easier solution, of course, is to get your milk in another form than the liquid form. Drink water or herb teas for beverages and get your calcium from cheese, yogurt, buttermilk, etc.

Not all people have the same degree of discomfort or react to the same amount of milk. **Some people have difficulty digesting only one glass of milk or less. Others can drink as much as four glasses before they have any discomfort.** However, according to an article in the *American Chemical Society Symposium Series, No. 15*, even those who have no noticeable symptoms when they drink milk may not absorb all its nutrients, because of their deficiency in this essential enzyme, lactase. And, as we have pointed out, the lactose is thus not absorbed. It stays in the intestine and is fermented to lactic acid which serves as a cathartic or purgative. Carbon dioxide and hydrogen are produced in the intestine by this fermentation, producing a frothy diarrhea.

Dr. Max Tesler, director of medical services of Biometric Testing and a member of the medical staffs of several New York City hospitals, tells us that the helpful bacteria called *Lactobacillus acidophilus* taken in or with plain sweet milk will allow someone intolerant of lactose to drink milk with no difficulty. Buttermilk, yogurt and cheese contain these helpful bacteria.

We have mentioned several times these helpful bacteria and how they function in the human colon. They are especially important for people who have been taking antibiotics by mouth, for these drugs destroy helpful as well as harmful bacteria. Diarrhea and other very troublesome disorders may result. *Moniliasis*, a fungus infection of the colon, may develop with "disastrous results," according to Dr. Tesler. We know from other experts that this kind of fungus infection can also get started in female reproductive organs (cervix and vagina) with, again, disastrous results.

This fungus tends always to produce an overgrowth unless it is held in check by the *Acidophilus* bacteria.

Dr. Tesler also mentioned a serious complication in patients with severe liver disease in whom toxic effects on the brain can be prevented by the presence of a "significant population" of *Lactobacillus acidophilus* in the intestinal tract.

Dr. Eugene Jolly, a former Deputy Director of the New Drug Branch of the Food and Drug Administration, says that acidophilus bacteria tend to be difficult to establish in the human colon, even those present in unpasteurized yogurt, which is the reason we must take yogurt every day or so if we would have the healthiest possible digestive tract.

Says Dr. Jolly, "a good *Lactobacillus acidophilus* product would certainly be a boon to our entire population; particularly people dependent on laxatives or those who complain constantly of just plain indigestion."

As we have reported, fermented milks containing these bacteria have been dietary staples since early times among many peoples. Yogurt was the product developed by the biologist Metchnikoff in 1908. He attributed the longevity and good health of numerous Balkan people to their daily use of yogurt. The Dutch people have fed their babies buttermilk for centuries to prevent infantile diarrhea. And early American doctors used this food for the same purpose. It establishes millions of live *Lactobacillus acidophilus* in the colon. These bacteria suppress the putrefactive, gas-forming and sometimes disease-causing bacteria.

Dr. Henry Isenberg, Chairman of the American Board of Medical Microbiology, Professor of Clinical Pathology at the State University of Stony Brook, New York, and Chairman of the Department of Microbiology at the Long Island Jewish-Hillside Medical Center, also New York, says that modern food processing and the addition of antibiotics to many foods, as well as their almost universal use as drugs, disrupt the delicate balance so essential for the proper functioning of digestion.

"Once established and maintained in its proper sphere of activity... *Lactobacillus acidophilus* can produce sufficient acid to limit not only the growth of the innumerable putrefacient microorganisms but also to interfere with the accumulation of their metabolic end products..."

These putrefactive end products challenge the liver's ability to deal with them because it is already overloaded with additives, preservatives and drugs which it must work hard to detoxify.

The constant presence of sufficient numbers of *Lactobacillus acidophilus* may prevent or diminish the opportunity for the establishment of bacteria known to cause dysentery and enteritis. They also prevent the establishment of bacteria which are resistant to antibiotics, hence capable of causing many disorders. Dr. Isenberg says that taking antibiotics by mouth is the commonest cause of persistent irritating itching "of those parts of our anatomy still considered 'unscratchable' in public." Itching anus and reproductive organs are the unmentionable and embarrassing complications of antibiotic drugs in millions of people. The wholly beneficial *Acidophilus* bacteria can dispose of such troubles speedily. There is also considerable evidence that these bacteria are very helpful in preventing "cold sores" or "fever blisters" (*herpes simplex* the doctors call them).

The answer seems to be unpasteurized yogurt eaten almost every day, or to drink buttermilk. For a time a new Acidophilus milk which tastes like regular milk was on the market. We do not know whether or not it is still on the market. Developed in North Carolina, it was being distributed by several dairies in 1977.

Milk, as produced by the cow, has a lower percentage of water than many succulent fruits and vegetables, we are told by the *Yearbook of Agriculture, 1959,* an annual publication of the U. S. Department of Agriculture which deals with various subjects in great detail each year. The 1959 edition is called *Food.*

"The average composition of milk is about 87 per cent

water, 3.5 per cent protein, 4 per cent fat, 5 per cent carbohydrate, and a little less than 1 per cent ash (minerals). These milk solids are an outstanding source of calcium and a good source of riboflavin (vitamin B2), high-quality protein, vitamin A, thiamine (vitamin B1), vitamin B12 and other dietary essentials," *Food* reports.

"A cream line forms in milk upon standing as fat particles rise to the top. Commercial separation of the cream from the milk is done so effectively that less than 0.1 per cent of fat is left in the skim milk."

The publication goes on to say that homogenized milk, which forms a major portion of the whole milk on the market, has no cream line. The fat globules are broken up by a mechanical process into such tiny particles that they no longer rise to the top. The nutrients are uniformly distributed in homogenized milk, as they are in evaporated milk.

Skim milk, fluid and powdered, has become increasingly important in the family grocery orders, the yearbook adds. Since nearly all the fat is removed from skim milk, its energy value is reduced greatly. A glass of skim milk has only half the calories of a glass of whole milk.

Removal of the cream removes the fat-soluble nutrients carried in the cream—vitamins A, D, E and K—the most important of which is vitamin A, *Food* reports. A pint of whole milk provides nearly a sixth of the entire daily allowance of vitamin A for an adult, but a pint of nonfat milk has only a negligible amount. The water-soluble nutrients, the minerals, the B vitamins and the protein originally present in the whole milk remain in the nonfat portion.

"Milk is the basic ingredient of many manufactured dairy products. Their nutrient content depends on whether the starting point was whole milk or one of the fractions (cream or skim milk) separated from it and whether there is any further separation of milk solids. A decrease or increase in nutrients may result from addition of other ingredients," *Food* says.

Buttermilk originally was the fluid portion left after separation of the butterfat. It had nearly all the solids (other than the fat) that were in the cream. Most of the fluid buttermilk on the market now is made directly from skim milk; a culture is used to develop the desired flavor and consistency. Cultured buttermilk has somewhat more lactic acid but otherwise has practically the same composition as the skim milk from which it is made, the USDA publication goes on.

"Special therapeutic properties are sometimes attributed to buttermilk and other fermented milks (kumiss, kefir, yogurt)," *Food* reports. "The protein precipitate in them is in the form of a fine curd, which may permit them to be digested more quickly than plain milk."

"The amounts of the main nutrients in the milk used in making the fermented products are not greatly changed. The same nutritive values may be applied. Yogurt, for example, can be made many ways—from skim milk, whole milk, evaporated milk or from mixtures of these. The nutrient content of the milk ingredients used applies to the prepared yogurt."

Pasteurization of raw milk is a necessary safeguard, *Food* says. It does not free milk entirely of bacteria, but it destroys those that cause diphtheria, tuberculosis, typhoid, undulant fever and other diseases. Loss of nutrients through pasteurization is insignificant compared to the safety it provides, the USDA reports.

"Pasteurization does not affect materially the main contribution of milk and milk products to the diet—that is, the calcium, protein, riboflavin and vitamin A. The losses induced by heating are chiefly in the vitamins ascorbic acid (vitamin C) and thiamine (Vitamin B1)—losses that easily can be made up in a diet composed of a good variety from several other groups of food."

The USDA yearbook does not mention it, but raw milk is available in some localities through health food stores. In some states you need a prescription to buy the raw milk; in others, it is available over-the-counter. Raw-milk dairies are usually subjected to far more stringent regulations than regular dairies. Your local health department can give you guidelines for purchasing raw milk in your locality, if it is available and you wish to do so.

Minerals and Trace Minerals in 100 Grams of Milk

(About ½ cup of liquid milk or cottage cheese)

Minerals

	milligrams
Calcium	144
Iron	0.057
Magnesium	13
Phosphorus	93
Potassium	144
Sodium	50

Trace minerals

	micrograms
Aluminum	350
Boron	60
Bromine	400
Chromium	1.4
Cobalt	0.13
Copper	32
Fluorine	30
Iodine	35
Manganese	5.5
Molybdenum	6
Nickel	6.5
Selenium	2.5
Silicon	82
Vanadium	Trace
Zinc	450

Milk and milk products provide about two-thirds of the total calcium in our diets, nearly half the vitamin B2 and more than one-fifth of the protein, *Food* reports. Calcium and protein are well retained in milk, it adds.

Vitamin B2 in milk is reduced by exposure to direct sunlight, daylight or artificial light, *Food* tells us. The rate of destruction is affected by the intensity of the light, length of exposure and the temperature of the milk.

The total loss of vitamin B2 from the time of production until the milk is served need not be large if it is handled properly—if it is kept clean and cold and out of direct sunlight.

"Milk under artificial light in refrigerated showcases loses little of its riboflavin," the USDA says. "The milk there is cold, and the light bulbs usually are of low intensity. If the showcase is near a window so that the milk is exposed to considerable daylight, however, losses could increase enough to be important."

"Milk delivered on the doorstep should be protected from light quickly. Within the first five minutes there is little loss of riboflavin from milk exposed to the sun in clear-glass quart bottles. By the end of 30 minutes, losses are about 10 per cent. They reach about 40 per cent after two hours, even when the temperature of the milk does not rise above 70°. The loss increases as the milk warms.

"The size and type of container are related to retention of riboflavin," *Food* continues. "Milk in half-pint bottles loses about twice as much as milk in 2-quart bottles. Paper containers and brown glass provide several times as much protection to the riboflavin as clear glass."

Is lumpy evaporated milk safe to use? The USDA states that lumps in evaporated milk are formed by the solids settling during storage. The lumps do not harm the milk. Cans of evaporated milk can be turned or shaken at frequent intervals during storage to prevent lumping.

Can you drink too much milk? Yes—if milk is consumed in such large amounts that it crowds out other important foods from the diet, the USDA says.

Have you often wondered why some milks have different colors? For example, the yellowness of milks of a Jersey cow, a Holstein cow, a ewe, a goat and a sow may be different, even though they are all grazing on the same green grass. The milk of the Jersey is yellower than that of the Holstein. The milks of the other three species are nearly white.

"The yellow color of milk and cream is due to carotene," *Food* reminds us. "Vitamin A is almost colorless. Animals with great ability to convert carotene to the vitamin produce white or slightly yellow milk. Those that are not so efficient put more carotene and less vitamin A into their milk."

"The total vitamin A value of milk, cream, butter and eggs is the sum of the vitamin A and the carotene present, but one cannot estimate the vitamin A value of such foods on the basis of their color alone."

Carotene, as you remember, is a pigment found in plants (carrots, etc.) which is transformed into vitamin A in the liver. Vitamin A by itself occurs only in foods of animal origin.

Writing in *Food*, Miriam E. Lowenberg, of Pennsylvania State University, said that some children find drinking even two cups of milk a day too much. Among a group of children up to five years old who were physically healthy and those whose patterns of drinking milk were observed, the amount of milk taken daily declined after age two. When the children became of school age, they again increased their consumption of milk. "We can expect that many children of these ages will drink less milk than they did earlier. If mothers expect such a development and allow then to drink as much as they want, with neither direct nor indirect forcing, the children will not become unfriendly toward milk as a food," she says.

"In my observations of the milk-drinking tendencies of thousands of children in nursery schools, I have found that the amount a child takes is related directly to the temperature of the milk," Ms. Lowenberg reports. "The child of two, three or often four years prefers his milk lukewarm and not icy cold.

"In two large nursery schools conducted for the children of workers in a shipyard during (World War II), the milk sometimes was served ice cold when the kitchen workers had not followed directions. Far less milk was drunk on those days. On days the supervisor checked carefully to be sure that the milk was at room temperature, each child drank eight ounces of milk at a meal without any urging.

"I have verified this in my work with mothers and in feeding children in their homes. I believe, therefore, that attention to temperature is important in giving milk to children two to six years old. I found that children had come to appreciate cold milk at about five or six years (sometimes even at four years)," she continues.

Another factor of seeming importance in serving milk to children in groups is the way in which the milk is offered to them. Small glasses that hold about three-fourths of a measuring cup of milk when full have been found to be best, Ms. Lowenberg says.

"These can be poured about two-thirds full. This gives only four ounces, or about one-half of a measuring cup of milk in the serving. One glass presented with the main part of the meal and one given with dessert may be taken readily by most of the children in the group. The small glasses are easily handled by small hands, and the goal looks possible to the young child, she adds.

For the child who cannot drink milk easily, he can, of course, be offered Cheddar-type cheese, cottage cheese and extra portions of meat, eggs, fish and poultry. "Certainly he should not be forced to drink milk, and the mother should not transfer to him her concern or worry about the lack of milk," Ms. Lowenberg states.

When the child is served slightly less than the adult thinks he will eat or drink, he has a chance to be successful. This rule also holds for all foods for most children. Children also like to set their own goals by pouring their milk from small pitchers, and they often drink more when they pour it themselves, Ms. Lowenberg believes.

She adds that forced feeding is or should be a dead issue. It leads to nowhere, she says.

Ms. Lowenberg reports that some milk can be "eaten" in drinkable soups, custards, puddings and cereals or vegetables cooked in milk. Nonfat dried milk can be added to many cooked foods. You can count about four tablespoons of dried milk as equal in nutritive value to a cup of skim milk. It can be used to fortify many foods, such as gravies, sauces, puddings and cooked starchy vegetables. Cheddar cheese and cottage cheese, often eaten by children who refuse to drink milk, add to the protein, calcium and riboflavin content of the day's food.

"Many times interruptions in drinking milk are just a temporary whim; if other foods are substituted for milk and no special attention is called to it, the child comes back to liking it again," she says.

Fresh dairy products should be kept cold and tightly wrapped or covered so that they do not absorb odors and flavors of other foods, *Food* reports. A storage temperature of 40° is most desirable in protecting flavor and food value of milk and cream.

As soon after purchase or delivery as possible, the glass bottle or paper carton should be rinsed under cold running water, dried and refrigerated promptly.

Evaporated and condensed milk may be stored at room temperature until the container is opened. Then it should be refrigerated in the same way as fresh fluid milk.

Dry milks will keep for several months at room temperature of 75° or lower, or they may be kept in the refrigerator. Nonfat dry milk is somewhat more stable than whole dry milk because of its lack of fat. Both should be stored in tightly covered containers to prevent moisture absorption, which causes off-flavors to develop and makes reconstitution difficult.

Variations in components of milk depend on the breed of

cow, the stage of lactation, the type of food the cow is eating, the season and the other physiological inherited and environmental factors, according to *Dairy Council Digest*, January-February, 1971. For instance, "Levels of certain minerals such as zinc, manganese, cobalt and iodine are affected by amounts in the diet of the cow, although other minerals such as iron are not appreciably influenced by dietary level."

This sentence demonstrates clearly the superiority of organically raised dairy herds, although the *Digest* does not comment on this. Obviously, when cows are fed from pasture which has been fertilized only with commercial fertilizer, they, and hence their milk, will probably be lacking in trace minerals which eventually disappear or decrease in the soil when the crops are taken off year after year without replacing these minerals by turning back organic matter to the soil.

Levels of the fat-soluble vitamins A, D and E in milk depend on the amount in the cow's diet. The B vitamins and vitamin E are produced in the cow's digestive tract, so they are not essential in the fodder with which the cow is fed.

"Market milk is made up of milk from many herds and is subject to standardization of fat and solids content to comply with legal regulations," says the *Digest*. "Thus milk that reaches the consumer is more of a standardized product than many other natural foods."

We would like to point out that this pooling of commercial milk from many sources guarantees us the best of each lot, but it also guarantees us that whatever one dairyman is using in his barn or in his feed will influence the wholesomeness of the entire lot. If every farmer in a given area obeys the law and withholds antibiotics from his cows for the required period before selling his milk, there will be no penicillin in that entire batch of milk to threaten any allergic individual who drinks it. But if all farmers except one obey the law, his milk will contain the antibiotic which will contaminate the entire batch. The same goes for any pesticides taken in by the cows in food and water. This is another reason for seeking out organically produced dairy products, if they can be found in your locality.

The protein of milk is of a very high grade, its biological value being 84.5 per cent—that is, almost all of the protein of milk is retained by the person drinking it. So far as digestibility goes, 96.9 per cent of the protein of milk is completely digestible. The fatty part of milk is very complex, containing many different kinds of saturated and unsaturated fats.

But won't you get fat, if you use much milk or other dairy products? Statistics have shown that dairy products nationally contribute only 11.3 per cent of the calories eaten by the American people, but contribute 75.8 per cent of the calcium, 42.4 per cent of the vitamin B2, 36 per cent of the phosphorus, 28.8 per cent of the magnesium, 22.1 per cent of the protein, 20.6 per cent of the vitamin B12 and so on. So comparing dairy products to almost any other foods, you find that the number of calories (the overweight producing element of foods) is very small compared to the wealth of nutrients contributed to the national diet by dairy products.

When milk is made into skim milk and cream, the fatty part is skimmed off and sold as cream. Most of the fat-soluble vitamins are thus not present in skim milk, but all other nutrients are there in about the same quantity as they are in whole milk. The same is true of light table cream. It has about the same amount of protein, vitamins and minerals as milk, but considerably more fat. It is unlikely that anyone will get too much of this fat, since one easily becomes satiated on fat. Any more produces nausea.

Some people have managed to cut butter fat out of their meals entirely, using no cream, no milk but skim milk, and using vegetable oils rather than butter. In this case, it is well to keep in mind that the vitamin A and the small amount of vitamin D present in butter fat must be supplied by some other item in the diet.

Did you know that a dairy goat or two can supply your

family with good, wholesome, naturally homogenized milk for much less money than it takes to keep a cow? It may or may not be the practical thing for you to do, but you may want to consider the idea if you have a place to keep them. Goats cost less to start with. A doe might cost around $50.

Goats need less space, pasture and feed than cows. They are also good scavengers, eating just about anything that isn't nailed down. And a good doe (female goat) produces two quarts of milk or more a day for about 10 months of the year. A top milker could yield three to five quarts a day for the same length of time.

Keep in mind, however, that a goat needs about half an acre of pasture during the grazing season of five or six months. Goats thrive on any good mixed pasture, but do not like clover by itself. What do they eat? Many kinds of food—beets, turnips, cabbages and carrots. They also need iodized or block salt at all times. You need to supply plenty of clean, fresh water. No special kind of housing is needed—a barn or shed that is dry and free from drafts will do. If you really are interested, check with your county (agricultural) agent or state university. The Agricultural Research Service at the U.S. Department of Agriculture has the figures on costs of having a dairy goat for home milk production. This information has been extracted from *Food and Home News*, U. S. Department of Agriculture, Washington, D. C. 20250.

Of course, if you live in an urban area, you should check with your local zoning board, township, city or county officials to make certain you are allowed to keep animals in the zoning area in which you live.

Remember, please, that milk contains everything that a human being needs for growth and survival except iron, which it lacks. In pasteurized milk its vitamin C content cannot be depended upon. But its protein content is superior to that of any other food except eggs, in that it contains all the amino acids in the best possible balance for economical, full nutrition. Its calcium content is our best source of this important mineral. It has an almost equal amount of phosphorus, calcium's twin mineral. It is rich in the B vitamins, especially the scarce one, riboflavin.

Don't neglect milk and milk products in meal planning, especially if you are cooking for children or old folks. Its long history of nourishing man recommends it as one of our finest foods.

CHAPTER 9

Yogurt

MILK, LEFT TO ITSELF, will ferment. It becomes acid and takes on a solid form. You then have curdled milk whose nutritive qualities have been known since antiquity.

Many different kinds of bacteria which live in milk can cause it to clot or curdle. The products formed during the fermentation process differ according to the different kinds of bacteria present. Therefore, one cannot obtain clotted or soured milk with the certainty that it will always react in some certain way on a certain body mechanism, unless one knows exactly what kind of bacteria have been placed in the milk.

Yogurt is a sour milk product obtained by inoculating the milk with souring agents, in the form of healthful, beneficial bacteria. It is prepared of whole milk, or milk from which the cream has been partially or completely removed, pasteurized, homogenized or not, condensed or not. This diversity in the kind of milk that can be used explains the difference in composition and dietary value that may exist in different kinds of yogurt, depending upon how they were made.

According to Persian tradition, an angel revealed to the prophet Abraham the method of preparing yogurt. To this food he owed his fecundity and longevity. He lived, the Bible tells us, to the age of 175. A similar cultured milk, kefir, or "the champagne of milk," was known in antiquity as the

"Drink of the Prophet." The substance used to prepare it was called the "Drink Grains of the Prophet Mohammed," since he introduced it into his religion.

Yogurt and kefir were made from milk of sheep, buffalo, goats, mares, cows and llamas. During ancient times the method of preparation was handed down as a precious inheritance from father to son in many countries. In France, we are told, the Emperor Francis I, who lived in the 16th century, had been ailing. Court doctors had exhausted their remedies. Someone at court knew of a famous doctor in Constantinople who had a "secret" medicine. He was sent for. He brought his own goats and concocted the "medicine" from the goat milk. The medicine was yogurt. Since that time, Frenchmen have called this tasty form of milk "Lait de la vie eternelle" (milk of eternal life).

Later, the success of this food in France came from the research (and not very scientific research) of Metchnikoff, who was a disciple of Pasteur (the father of pasteurization). He drew attention to soured milk because the long life of people who lived in the Balkan countries seemed to be related to their consumption of this food. . . . The particular bacteria which are present in yogurt would grow, he thought, in the intestinal tract, modifying the nature of the bacteria there and the composition of the intestinal climate or background. According to this theory, the acid fermentations which come into being when yogurt is made live and grow in the intestine, do away with putrefaction and neutralize the alkaline products of this putrefaction, we are told in an article translated from an article by Yvonne Depuis, Pierre Brun, Paul Fournier and Denise Risch in a French journal, *L'Alimentation et La Vie.*

While milk is fermenting to form yogurt, certain organic acids are formed, especially lactic acid, the article says. This is what gives the yogurt its pleasant acid taste. But because there are acids present in the yogurt, a theory was developed (a not very intelligent theory, the authors believe) that yogurt is demineralizing or decalcifying. "It is possible, of course, to take the minerals out of a bone, by crushing it and mixing it with an acid. So there was some question as to whether the organic acids of yogurt neutralize the alkalinity of the calcium of bone. Metabolic researches seemed to disprove this theory. Milk is certainly the food which causes bones to ossify, or become mineralized and strong. It is also an abundant source of calcium phosphate, the most important mineral in bone formation. But, precisely because the mineral elements of milk (notably calcium) are especially well used by the body, (1) therefore, milk efficiently protects the mother's skeleton during lactation (breast-feeding) and (2) this protective effect where the skeleton is concerned is brought about by milk sugar or lactose."

"In the present experiment," the article reports, "we observe the influence of yogurt on the way the body uses calcium. The subject is not new. The composition of yogurt and its nutritive value have been the subject of important researches. Among the most recent of these is a test to determine its vitamin content, and a study to determine the relative utilization by the body of the calcium of milk and that of yogurt, showing the advantage of the latter as a source of calcium . . .

"A recent series of experiments, using young rats, showed that, much more important than the quantity or the form of calcium eaten by the animal is the presence in the diet of certain elements which regulate the use of calcium by the animal. Among the tests of a satisfactory use of calcium is this: that the level of calcium in the blood must be maintained at its normal level. The young rat, receiving a balanced diet, with the normal amount of calcium, a diet in which the relation of calcium to phosphorus is perfect, cannot maintain the proper use of calcium by the body unless the diet also contains vitamin D and either lactose or other elements closely allied to lactose.

"**Yogurt is, without doubt, an abundant source of calcium.** And it also contains potent factors for guaranteeing the proper use of that calcium; lactose and, if the yogurt is prepared from whole milk, vitamin D. That is why we decided to study the effectiveness of yogurt as a calcifying agent. We used very sensitive tests of the levels of calcium in relation to lactose and vitamin D."

"The experiment shows that yogurt is extremely effective in producing the best possible use of calcium by the body. The presence of acid in yogurt does not prevent the vitamin D and the lactose from exerting their powerful calcifying activity to help the body to use the calcium. . . . From the time yogurt was introduced into their diet, the rats absorbed, retained and used an ever-increasing amount of calcium. The level of calcium in their blood remained normal. An improvement in the utilization of calcium while eating yogurt was observed in all the animals. This result was remarkably uniform. The more yogurt eaten, the better the results," the article concluded.

It seems there is a sugar in milk called *galactose*. Just as there are some human beings who lack the enzyme to deal with another milk sugar, *lactose*, there are some people who lack the enzyme *galactokinase*, so they cannot and should not eat much galactose. It gives them troubles. And one of the troubles is likely to be cataracts, the filmy opacity of the lens of the eye or the capsule in which the lens is placed.

Other people—a very small group—suffer from an inherited disorder called *galactosemia* or galactose diabetes. These people do not have in their bodies another enzyme by the name of *galactose 1-phosphate uridyl transferase* which is necessary to convert galactose into glucose which the body can use. These people born with this defect, do not thrive in infancy, they get jaundice, they may have troubles with liver and spleen, they may be mentally defective—and they may have cataracts.

None of these disturbing disorders pose any threat to the rest of us who have no trouble with absorbing and using this special kind of sugar. But researchers have known for about 30 years that rats—the rats used in laboratory experiments—all lack the enzyme necessary for using this milk sugar correctly. So if they are fed a diet in which there are large amounts of this sugar they will almost certainly get cataracts.

In 1970, two Johns Hopkins scientists decided to try an experiment in which they would feed rats a diet consisting of nothing but yogurt. Rats make their own vitamin C, so presumably such a diet could sustain the animals in good health throughout their lives.

Dr. Curt Richter and Dr. James R. Duke used commercial yogurt on purpose, rather than making it in the laboratory, because most commercial yogurt is made from skim milk and is enriched with added powdered skim milk. The powdered milk, too, contains almost no fat. So modern commercial yogurt is usually advertised as a great food for a reducing diet because of its low fat content.

But, when you remove any ingredient of a food, you concentrate what is left in the food. Whole milk, as it comes from the cow, is about 4 per cent fat. In powdered whole milk, you have removed the water, so the fat is concentrated and powdered whole milk is about 28 per cent fat. By making yogurt of skim milk and adding powdered skim milk, the yogurt industry concentrates other ingredients, including galactose. So commercial yogurt would naturally contain more galactose per cup than would homemade yogurt made from whole milk or whole milk with powdered whole milk added.

So the rats in the Johns Hopkins laboratory wre getting a whopping dose of galactose every meal, every day. That was what the scientists planned. They believed that the rats would eventually get cataracts from eating nothing at all but this high-galactose yogurt throughout their lifetimes. And sure enough, they did. They got the cataracts, and because all the rats are unable to eat galactose without getting cataracts, and because they got nothing to eat except for the yogurt. So all the dice were purposely loaded against these rats. However, they lived long lives in good health and suffered only from cataracts which, we suppose, would not trouble them very much since laboratory rats have little need for sharp eyesight.

However, when it comes to human beings, it's a different story. All of us want to preserve our good eyesight as long as possible. All of us want desperately to avoid cataracts. So the

Food Value of 100 grams of Yogurt (Less than ½ cup)

Water		87.10 grams
Dry material		12.90 grams
Protein		4.22 grams
Total fat		1.93 grams
Lactose (milk sugar)		5.45 grams
Minerals:		
Calcium		.174 grams
Phosphorus		.125 grams
Sodium		.023 grams
Potassium		.066 grams
Vitamins:		
Vitamin A		95. I.U.
Vitamin E		.62 milligrams
Vitamin B1	(Thiamine)	.235 milligrams
Vitamin B2	(Riboflavin)	.227 milligrams
Vitamin B6	(Pyridoxine)	.0176 milligrams
Vitamin B12		Traces
Niacin		.104 milligrams
Pantothenic Acid		.313 milligrams
Essential Amino Acids (Proteins)		
L-Arginine		.123 grams
L-Cystine		.080 grams
L-Histidine		.088 grams
L-Isoleucine		.267 grams
L-Leucine		.571 grams
L-Methionine		.129 grams
L-Phenylalanine		.139 grams
L-Threonine		.407 grams
L-Tryptophane		.052 grams
L-Valine		.574 grams

Johns Hopkins doctors rapidly became the center of boiling controversy. Letters poured in from people who did not know the details of the experiment, but had read only the newspaper headlines which said, "Rats get cataracts from eating yogurt."

Hastily the researchers prepared an article which was published in the *Journal of the American Medical Association* for December 7, 1970, to clarify what they had done and why. In this article they pointed out what we have discussed above--the nature of the diet fed to these rats and the fact that all rats inherit an inability to deal with galactose. It was as if some almighty experimenter had taken a group of people, all violently allergic to strawberries, let's say, and fed them nothing but strawberries all their lives.

So the rats got cataracts, just as the doctors knew they would. But—and it is an important but—the researchers were earnestly trying to show something else which they hoped would be helpful to people. They quote a British experiment done in 1939 in which rats were fed lots of galactose and lots of fat at the same time. Not a single rat got cataracts. No matter how hard they tried, these British scientists could not produce a single cataract so long as they gave the rats plenty of fat.

The Johns Hopkins rats were getting almost no fat, remember. We have no inside information, but it seems likely that the Baltimore scientists will next try an all-yogurt diet made from whole milk, or will use commercial yogurt and add some butter or oil. Then we'll see whether the cataracts are prevented.

Drs. Richter and Duke tell us that cataracts are quite common in India. In some parts of India yogurt makes up almost all the diet. And this is made from skim milk, so it is very low in fat. It's possible, the scientists believe, that this kind of yogurt, eaten as almost the only item of diet, may be causing some of the cataracts.

So the point they wanted to stress is made near the end of their *JAMA* article. They say, "Milk is one of the very few substances that has been produced exclusively as a food—chiefly for the young. During the 75 million years since the appearance of mammals, evolutionary processes have worked out in the most precise way the proportions of the various ingredients in milk. For most species of mammals, these ingredients and proportions are much the same. Perhaps it is inadvisable to alter these proportions or add to these ingredients that have been worked out with such great care by nature."

In other words, maybe it's very unwise to go on a low-fat diet and cut out entirely or cut way down on the purely natural fat contained in milk. It's there for a purpose. Nature put it there for some very sound reasons apparently. Because of our panic over the word "cholesterol" and our terrible problems of trying to avoid overweight, we may be doing ourselves harm by using only skim milk, if we depend on milk for much of our food.

We know very little about the milk sugar, galactose, say the Johns Hopkins men. It is not very well assimilated and it appears to have little nutritive value. But, in one study done in 1948, Dr. Richter discovered that one purpose of galactose is to help the body use fats correctly. So perhaps the worst thing we could do is exactly what these scientists did to their rats—to give them large amounts of this sugar whose job it is to combine with fats and then remove all the fats from the diet, so that there is nothing for the sugar to combine with! Could this be the reason for the cataracts? Remember they do not appear when there is fat in the diet along with the galactose.

"Much research on this fascinating sugar remains to be done," say the scientists. And they have this to say about yogurt—that is, the traditional yogurt made from whole milk. "The original yogurt recommended by Metchnikoff for a long, active, healthy life, is undoubtedly one of the best of all foods known to man. It has long served as the chief source of nourishment for many people and is one of the safest of all foods in regard to contamination. Archeologist friends of ours have told us that when traveling in undeveloped areas of the world, they have found that the natural yogurt is always nutritious and safe to eat."

So, should you be afraid to eat yogurt? If you suffer from galactosemia or if you were born without the enzyme galactokinase, then you are undoubtedly on a diet low in galactose and you probably avoid all dairy products. You are thus a very special person with a very special dietary problem. Even so, it seems unlikely that you will get cataracts.

For the rest of us, without this inherited disability, there is absolutely not a shred of evidence that eating yogurt—even commercial yogurt made with skim milk and without a bit of fat in it—will cause cataracts, unless we should, for some foolish reason, decide to eat nothing but yogurt for an entire lifetime. In that case, we might or might not get a cataract. But surely none of us is contemplating such an unwise way of eating. Yogurt, as Drs. Richter and Duke pointed out, is a great food, a highly nutritious food whether made with or without fat. But it seems that whole milk yogurt may be preferable for all of us, since nature made milk with some fat in it and put a sugar into the milk whose job it is to combine with the fat.

In case you should decide to go on eating non-fat yogurt, there is still no possiblity of harm, since you will, of course, get fat in other foods. It's impossible to design a diet that is entirely fat-free and there is no reason why anyone should. Fat is an important part of a nourishing diet. We need it to assimilate all fat-soluble vitamins—A, D, E and K. We need it to keep skin and hair soft and pliable. It has many other uses in the body.

If you want to reduce, it is our earnest recommendation that you concentrate on eliminating all refined carbohydrates from your meals, cut down on salt, cut down on drugs like coffee, alcohol and cigarettes and stop worrying about fats. If you want to make your own yogurt at home—and lots of people do—your health food store has all the makings. It's fun. You can make it with whole milk or skim milk. And enjoy it.

Would you believe that yogurt bacteria have been found to protect against cancer? True. *The Journal of the National Cancer Institute* published several years ago the account of an experiment at the University of Nebraska, where two groups of mice were given the same diet, but one group got some yogurt in addition. Both groups were then infected with cancer cells.

After eight days an average of 28 per cent inhibition of cell proliferation was evident in the mice given yogurt. That is, 28 per cent of the tumor cells did not grow in the mice which got just a bit of yogurt. In the mice which did not get the yogurt, the tumor cells continued to grow as expected.

The scientists, reporting this astonishing fact, said, "The above findings suggest that lactobacillic cultures synthesize components which have an anti-tumor effect." They do not know what these components are and they plan further experiments to find out. It would apparently be quite impossible for them to recommend, on the basis of their experiment, that the general public, exposed on all sides to cancer-causing chemicals, eat yogurt regularly as a possible preventive. Such a suggestion would probably be looked upon as "faddist." But the National Cancer Institute seemed to feel that the experiment was significant enough to print.

There is considerable evidence from scientists in Europe, especially Eastern Europe, that extracts from yogurt effectively inhibit the growth of two kinds of cancer. This is in addition to the abundant evidence from many sources showing that taking yogurt every day creates healthful conditions in the digestive tract, so that harmful bacteria are discouraged from growing there and helpful ones are encouraged.

Digestive disorders plague almost everyone these days. Cancer of the colon is second only to lung cancer as a cause of disease and death. Isn't it just possible that helpful lactobacillus in yogurt might be able to lessen this toll a bit, might establish such a healthy environment in the average

American's colon that we could wipe out this epidemic of cancer? Only, of course, if the helpful bacteria of the yogurt culture remain in the yogurt when we eat it!

A number of medical researchers have reported on using yogurt as a topical treatment for various sores, infections, ulcers and cancers. In a 1969 *Journal of the American Medical Association*, a physician inquired about the possibility of applying dried yogurt as treatment for breast cancer. A representative of the American Cancer Society replied, "I suspect information about this technique is scarce because it is usually performed by nurses, both to help heal ulcerations and to dispel odors. According to our nursing consultant . . . the rationale for use of yogurt is to fight bacterial fermentation and growth, and bacteria flourish in an alkaline medium; yogurt produces an acid medium which prevents the growth of odor-forming bacteria.

"Yogurt and buttermilk may be used almost interchangeably," he continued, "except yogurt flakes are more expensive and so are usually restricted to small or virtually inaccessible areas. For example, the flakes would be appropriate for head and neck ulcerations, while buttermilk might be used for a large decubitus ulcer (bed sore). Buttermilk is effective in irrigations, such as vaginal douches or colostomy irrigations. Yogurt and buttermilk are in daily use at Calvary Hospital in the Bronx, New York, a facility for the care of patients with advanced cancer."

This statement seems to illustrate the tendency for doctors to permit the use of just about anything, orthodox or unorthodox, when it comes to treating a serious or fatal disorder, but to scoff at the idea of preventing ailments with the same treatment. How many human ailments are related to bacteria of the harmful kind? The list is almost endless. How many of these ailments get their start from harmful bacteria that inhabit the human intestine? There must be many.

When the health seeker manages to control these harmful intestinal bacteria with yogurt or buttermilk, which encourage the friendly or helpful bacteria, the medical profession declares with one voice that this is imaginary nonsense. A myth. A come-on. A fraud.

But when a terminal cancer patient is afflicted with a wound so terrible that the odor from the bacteria inhabiting it is intolerable, what is used by the nurses to overcome these bacteria and establish the healthful acid surroundings that are needed? Yogurt and buttermilk, applied directly to the wound.

Why should we not make use of these same foods for controlling other bacteria that molest us? We have heard from several sources that taking yogurt or lactobacillus tablets every day and avoiding sugar will clear up acne "bumps" almost at once. They return speedily when you go back to the sugary goodies and the yogurt is forgotten. Could we not wipe out acne with a sugar-free, yogurt-rich diet? What about cold sores? What about the bacterial ramifications that sometimes follow colds? Why not try wiping them out, if possible, with this wholly beneficial, delightfully flavorful food?

New evidence is emerging from a Vanderbilt University study of the effects of yogurt on levels of blood cholesterol. The story is staggering in its implications for the health of those Americans who are desperately trying to lower high blood levels of cholesterol.

It all started when a Vanderbilt University specialist, Dr. George Mann, took the assignment of testing a food additive for its effect on blood cholesterol. The chemical is a surfactant, a material which makes it easier for food technologists to mix oil and water to produce gloriously smooth ice cream, mayonnaise, chocolate and so on. Surfactants are also generally used in detergents.

Experiments with mice had shown that when the surfactants were added to their chow, blood cholesterol levels went up. Dr. Mann decided to try the same experiment with human beings, but was faced with the fact that he probably could not find any Americans who weren't already eating lots of these substances in their food, so there would be no "controls" for the experiment—that is, people who had never eaten these chemicals.

He decided to go to Africa and work with the Masai, a nation of Africans who live completely on the products of their cattle—milk, blood and, infrequently, meat. Such people had obviously never tasted a surfactant. In addition, their blood cholesterol levels are very low, considering that they eat lots of fat every day in the whole milk which they drink or eat in sour or clabbered form. Heart attacks and other circulatory disorders are almost unknown among them.

Dr. Mann went to Africa and assembled 24 Masai men who agreed to participate in his experiment. He fed them yogurt every day, and nothing else. Half of the Masai were also given some of the chemical surfactant which Dr. Mann was testing. They could eat all the yogurt they wanted. And they did. They ate more and more yogurt until they were finally eating 1.7 gallons of the white creamy stuff every day. Some of them began to put on weight. Dr. Mann knew that overweight usually raises blood cholesterol, so he wanted to prevent this, if possible.

When the experiment had progressed for only three weeks, the Masai got word that all was not well at home. Neighboring tribes were moving in on the Masai territory. The volunteers told Dr. Mann that they had to go home. So he reluctantly terminated the experiment, and brought back to Vanderbilt University the blood samples which he had taken throughout the test.

Back in his laboratory, he was amazed to discover that the levels of blood cholesterol of the Masai had gone steadily down while they were eating the yogurt. Those who were eating the most yogurt showed the greatest decrease in blood cholesterol. And even those who had put on most weight showed a decrease in their blood levels of this fatty substance, cholesterol, which is believed to contribute to hardening of the arteries, strokes and heart attack.

It seemed incredible, but apparently something in the yogurt was decreasing the amount of cholesterol in the body. And, incidentally, the surfactant chemical which he was testing made not the slightest bit of difference in cholesterol counts.

Dr. Mann immediately assembled a group of volunteers at Vanderbilt University and began to feed them yogurt. He took regular counts of their blood cholesterol. He found that, if a minimum of 3½ pints of yogurt are eaten daily, cholesterol levels will go down, in Americans just as they do

Here Are the International Names of Yogurt and Kefir

Yogurt Is or Was Called:

Laban raid	in Egypt
Lebeny	in Assyria
Jazmia	by the Tartars
Kesch	in Turkestan
Skyr	in Iceland
Kyael meelk	in Norway
Taete	in Scandinavia
Glumse	in Finland
Taetioc	in Lapland
Hangop	in Holland
Pumpermilch	in Germany
Huslanka	in Carpathia
Urgutrik	in Bohemia
Kunney	in Mongolia

Kefir Is or Was Called:

Kisla-varenyka	in the Caucasus
Gioddon	in Corsica
Katke	by the Tartars
Mazum	in Armenia
Ojran	in Greece
Kumys	in Romania

in the African Masai. Dr. Mann does not know why this is so. He thinks that the helpful bacteria in the yogurt—the lactobacillus—may produce a substance which stops the body's own production of cholesterol in the liver.

Generally speaking, the body has ways of regulating its cholesterol levels. When little fat is eaten and not much food containing cholesterol, like liver and eggs, the body makes its own cholesterol so that cholesterol in blood is kept at a suitable level. But when fatty food is eaten (and the whole milk from which yogurt is made is such a food) then the human being should manufacture less and less cholesterol so that the fatty substance is kept at a healthful level.

But many Americans, apparently, have lost this regulating mechanism so that, the more cholesterol they eat, the higher their blood levels become. Experts are not certain, but they believe that an excess of the fatty material piles up in arteries and clogs them, leading to hardening of the arteries, with the possibility of strokes or heart attacks.

Whether or not Dr. Mann finally discovers the magical substance in yogurt that helps the body to regulate its cholesterol levels, there is no reason why we health seekers should not use whatever discoveries he has made up to now. Yogurt is fine food, containing all the complete protein, the calcium and the B vitamins in whole milk. In addition, yogurt contains very special kinds of helpful bacteria which, as we have reported, keep human intestinal tracts healthy by overpowering the unhealthful intestinal bacteria, thus creating the best possible environment for the digestion and absorption of food, as well as regulating elimination.

Is it necessary to gulp down 3½ pints of yogurt every day to get all these benefits, as well as the additional one of controlled cholesterol levels? Apparently it is necessary. But this shouldn't be too hard an obstacle to overcome. Two cups of yogurt at every meal, plus another cup before bedtime is not a burden on one's digestion. It will not make you gain weight, since you are very likely to cut down on other foods, as you increase your intake of yogurt.

Then, too, there are other ways to get the bacteria which make yogurt so powerful. Your health food store has many products which contain the yogurt bacteria in tablet or powdered or liquid form. You take them with milk or fruit juice to provide the lactose or natural sugar on which the bacteria feed. So you might decide to eat yogurt at only one meal or two, and take some of these other forms of yogurt in addition. If you don't like the taste of yogurt—and it's easy to develop this taste—then the tablets or powder or liquid are the perfect answer.

Most people love the taste of yogurt as they do sour cream. You can eat it with chopped raw vegetables as a salad. You can eat it with fruit, or dried fruit, nuts, peanuts or berries as a delectable dessert. But mostly keep in mind Dr. Mann's experiment and eat plenty of yogurt.

Do you have to buy all this yogurt every day? Not at all. Make your own. Start with a yogurt maker, available at your health food store. This makes the job very easy. You will be successful in producing just the quality of yogurt you want every time you make it. To get the first batch started, use some yogurt from the health food store, or buy yogurt "starter." This is nothing more than a collection of the necessary bacteria. You stir it into the warm milk, so that the bacteria get a good start. Then you keep the yogurt at just the right temperature until the milk has thickened and the bacteria have become a large colony pervading every bit of milk. Then you eat. After the yogurt is made, keep it refrigerated so that it does not spoil.

How long can you keep plain yogurt in your refrigerator? "Yogurt has always been a fermented milk product containing billions of viable bacteria per gram," reported *Food Technology*, November, 1975. "An investigation of commercial yogurt samples revealed the following ranges of total plate counts (of bacteria) per gram: 26 million to 4,159 million at the second day after purchase; 700 million to 30,300 million at the tenth day; and 1,200 million to 71,700 million at the 20th day. In all samples, the counts increased from day 2 to day 10 when held at about 45 degrees F. in a walk-in refrigerator. In all samples except three, the counts continued to increase over the next 10 days of storage."

Of course, you might have predicted it. **Yogurt is one of the few foods left to us in the supermarket which you can feel absolutely safe in buying, if you stick to the unchemicalized, unsugared, unfruited, unflavored kinds.** Yogurt can and should consist of nothing but milk—either whole or skimmed with perhaps some powdered milk added—but, in any case nothing but milk.

PLUS—and this is the plus that makes the product—certain very healthful bacteria, mostly the *lactobacillus acidophilus* or *lactobacillus Bulgaricus*, which act very helpfully in the human intestine. These health-giving bacteria—and there are billions of them in every tablespoon of yogurt—overcome all the pathogenic or harmful bacteria and help the human colon to achieve perfection.

Perfection, in this case, means smooth, trouble-free handling of waste material, so that we are scarcely aware of the function. It means correction of either constipation or diarrhea. It means rather rapid transit of wastes through the colon, so that they do not harden and become almost impossible to move. This rapid transit is also believed to help eradicate danger of colon cancer when we eat foods containing cancer-causing elements, as all of us must in the world in which we live.

The bacteria used to make yogurt give it the pleasant acid taste which most people find so appetizing. In recent years, mostly because of the enthusiasm of the health food movement, yogurt has become a most popular supermarket food. The previously mentioned article in *Food Technology* tells us that sales of yogurt amounted to $125 million in 1974 and are undoubtedly much higher by now. The future for the yogurt industry looks bright, say the authors from Penn State University, Division of Food Science and Industry. So they did a survey to see just how many people eat yogurt and what they think of it in general.

The majority preferred fruit-flavored yogurt, they report, and most of them like the fruit all mixed up with the yogurt rather than at the bottom of the container. The surveyors asked the people they interviewed what they felt about "natural" yogurt. Over two-thirds said they prefer "natural" yogurt to one which contains synthetic additives, flavorings, dyes, preservatives and stabilizers.

And what about the bacteria it contains? Only 44 per cent of all the people questioned know that yogurt contains bacteria. The rest eat yogurt, presumably, because they like it, because it's a "fun" thing and very fashionable, and because of its excellent nutritional value. So, with their usual genius for going at things from the wrong end, food companies have sold the American public a "new" taste treat—yogurt. With goodness knows how much money spent on advertising, they have convinced a large segment of the population that they should eat yogurt.

Well and good. Nothing could make us happier. But now it seems—and why should it surprise us?—that the yogurt industry has decided to sell yogurt without bacteria. Believe it or not, that is the theme of this article in a trade magazine. By pasteurizing the yogurt at low heat, the manufacturers can retain the acid, tart taste but destroy all the bacteria so the product will keep for weeks or longer on supermarket shelves.

"As a matter of fact," says *Food Technology*, "both pasteurized and sterilized yogurt, with a much-prolonged shelf life—are already available. One major national brand is a pasteurized yogurt containing either no or relatively low numbers of bacteria. The question arises whether a consumer will be disappointed or feel cheated because the traditional expectation of massive numbers of bacteria was not satisfied by the pasteurized yogurt."

Why shouldn't they feel cheated?

One reason is that 56 per cent of yogurt-eaters studied apparently do not know that yogurt contains beneficial bacteria. They have been trained to believe that bacteria of any kind are "bad," so, presumably, any yogurt which boasts of having no bacteria should be the best to buy! But, aside from the protein, vitamins and minerals (the same as those of milk) and the sour taste, there is no reason to buy yogurt from which all the bacteria have been removed. You might as well buy plain milk for a lot less money.

The Powers That Be in Washington have a dilemma on their hands. They must, it seems, set up a "standard" for yogurt. For instance, they could rule that only yogurt containing the helpful bacteria can be labeled yogurt. Pasteurized yogurt would then have to be called artificial yogurt or something of the sort, which would supposedly warn the customer seeking helpful bacteria away from the product.

But most dairy products are consumed fairly near the place they are manufactured, because they are perishable. FDA regulations govern only products that cross state lines. So locally made yogurt would not have to use the FDA regulations. Unless state food laws demanded it, pasteurized yogurt would not have to be labeled as such. So the customer, seeking the helpful bacteria of natural, unpasteurized yogurt, would not be able to tell whether the product on the shelf contained them or not.

The trick, of course, comes in the definition of yogurt. Says *Food Technology*, "One traditional view going back through the medical folklore of the last century and way back to Biblical times is that yogurt has therapeutic properties and that the presence of viable (living) bacteria is responsible for this benefit. The subject is still a controversial one since there is simply too much myth surrounding yogurt and too little scientific evidence to support the view."

It's true, of course, that making yogurt goes back to Biblical times and that, in many remote primitive countries today, a daily swig of a pint or more of yogurt is credited with achieving superlative health and longevity. It is also true that modern medical literature, especially in Eastern Europe, contains considerable reportage of carefully controlled laboratory experiments showing that yogurt bacteria are indeed helpful against many kinds of bacterial invasion. Certain Eastern European scientists are regularly treating cancer with yogurt, we understand.

The Food and Drug Administration, which will make the final decision, is chronically of the opinion that nothing in food has any relation to health, good or bad, except for certain well delineated vitamin and mineral deficiencies which do not exist in the United States. So we can expect the FDA, acting in their traditional way, to declare that the bacteria in yogurt serve no purpose except to make the milk taste sour, so there is no need to indicate whether the stuff has been heated to destroy the bacteria. When that time comes, you, the health seeker, will have no choice but to make yogurt at home.

For the present, we strongly recommend that you buy your yogurt at the health food store where its valuable store of helpful bacteria is appreciated. If you must buy it at the supermarket, shun the "prettied up," flavored, dyed and fruited concoctions. These are bound to include chemical additives you don't want to eat and the addition of fruit accomplishes nothing but increasing the price. Why not buy plain, unflavored yogurt and add your own fruit at home? Or add chopped vegetables—chives, onions, carrots, parsley, nuts, sunflower seeds, etc.

Read labels. Don't buy pasteurized yogurt. It's cheaper to buy milk. Better still, as a substitute for yogurt, buy buttermilk. If you are in doubt as to whether or not the yogurt is pasteurized, write or phone the people who make it and ask them if their yogurt is pasteurized or if it still contains the helpful bacteria you are seeking. Their address is on the container. If you need more details, ask the librarian at your local library who can probably give you a more complete address.

CHAPTER 10

Cheese

ONCE UPON A TIME, according to legend, an Arabian merchant, Kanana, put his supply of camel's milk in a pouch made of sheep stomach and set off across the desert on a long day's journey. The heat and the rennet in the pouch caused the milk to separate into curds and whey by nightfall. The whey, being liquid, satisfied his thirst and the cheese, or curd, his hunger. This was the beginning of cheese, so to speak.

Today, the U. S. cheese industry is a growing industry. **Within the next few years, cheese may replace butter as the largest manufacturing outlet for milk.** There are more than 400 known cheese varieties. In a recent year, the average American ate more than 12 pounds of cheese, half of which was Cheddar, the nation's favorite.

Natural cheese is made by coagulating milk (as the Arabian did in his pouch), then separating the whey, or watery part, from the curd or solid part. Some natural cheeses are then "ripened" to develop their characteristic flavor and texture. Others are used unripened. Cheddar cheese may be labeled mild, medium, mellow, aged, sharp or very sharp. **Many people—health seekers among them—prefer natural cheeses to other forms of cheese because each natural cheese has its own characteristic flavor and texture.** Flavors range from bland cottage cheese to tangy

Blue or pungent Limburger. Textures vary from the smooth creaminess of cream cheese to the firm elasticity of Swiss.

Of much more recent development are the pasteurized process cheeses which have been melted, pasteurized, then mixed with an emulsifier to produce what can best be described as flavored cardboard. This is the kind of cheese

Nutrients in One Ounce of Cheese

Cheese	Calories	Protein, grams	Carbohydrate, grams	Fat, grams	*% of Recommended Daily Allowance*			
					Protein	Vitamin A	Riboflavin	Calcium
Cheddar	110	7	1	9	15%	4%	6%	20%
Swiss	100	8	0	8	15%	4%	4%	25%
Monterey	100	6	1	8	15%	4%	6%	15%
Mozzarella	90	7	1	6	15%	2%	4%	20%
Parmesan	110	10	1	7	20%	4%	4%	30%
Romano	100	9	1	7	20%	4%	4%	25%
Edam or Gouda	100	7	1	8	15%	4%	4%	15%
Blue	100	6	1	8	10%	4%	4%	15%
Provolone	90	7	1	7	15%	4%	4%	15%
Cream cheese	100	2	1	10	4%	2%	2%	2%
Cottage								
creamed	25	4	2.9	4.2	4%	0	2%	3%
uncreamed	23	4	2.7	0	4%	0	2%	3%

that comes already sliced and packaged in plastic. There is no rind or waste. They are "handy," "convenient." They melt easily, so these are the ones usually found in cheese sandwiches at restaurants.

Pasteurized processed cheese "food" is a blend of cheeses which may include other ingredients such as fruits, meats or spices. It contains less cheese and less fat and melts more quickly than regular processed cheese. Cheese "spread" has even more moisture, plus chemicals to prevent separation of the ingredients at room temperature.

"Cold pack cheese" and "cold pack cheese food" are American products sometimes called "potted cheese." They consist of mixtures of cheeses plus vinegar and spices. Other ingredients are also used. Cottage cheese can be called pot cheese or Schmierkase.

The Dairy Council Digest for May-June, 1975 gives us some valuable information on the nutritional goodies cheese offers, as well as recent analyses of its cholesterol content, and other fats, along with protein, carbohydrates and several vitamins and minerals.

Generally speaking, the cholesterol content of cheese depends upon the fat content. The fattiest cheeses have the most cholesterol. Cream cheese has the most since it is made, after all, of cream, rather than milk. Uncreamed cottage cheese has the least since it is made of skimmed milk and is the standby of all dieters. Cheddar cheeses of all kinds have more fat than Swiss. Mozzarella and Parmesan are rather low in fat, hence in cholesterol.

The protein content of cheese, as we know, is high. It is, in essence, concentrated milk—that is, milk with the water removed. Of course, there is much more to the process of making cheese than that. The *Digest* defines cheese as "the fresh or matured product obtained by draining after coagulation of milk, cream, skimmed or partially skimmed milk, buttermilk or a combination of some or all of these."

As we mentioned, **there are more than 400 varieties of cheese in the world with as many as 2,000 names.** They are classified on the basis of moisture, fat and calcium, age, texture or general appearance, type of milk used, type of ripening agent used and country of origin. The lower the moisture content, the firmer the cheese naturally. The slower the ripening process, the milder the flavor and the longer the keeping quality. Parmesan and Romano, Cheddar and Swiss are four examples of hard cheese. Limburger, Roquefort, Blue and Gorgonzola are semi-soft. Soft cheeses include Camembert and Brie as well as cottage and cream cheese.

When the liquid part is drained away, in making cheese, this becomes the whey which, for some incomprehensible reason, modern technological geniuses mostly waste. It is fed to stock animals and made into industrial products instead of being carefully saved and made into a highly nourishing food, which it might be, since its calcium and vitamin B content is very high. The whey is often used as an additive in processed products, to add to texture and nutritive value.

Protein is our most expensive and valuable food element without which we cannot get along healthfully. We must have a given amount of it daily to nourish cells and to replace worn out or injured ones. Only one ounce (28 grams) of cheese provides the amount of protein listed on the accompanying chart, with some of the Italian cheeses ranking highest and the soft cheeses having the least protein, partly because they contain more moisture.

As we have learned, calcium is the mineral which is most valuable to us in cheese. **Dairy products in general are the best source of this essential mineral which keeps our bones and teeth healthy as well as contributing to the action of many other body mechanisms.** As you can see from the chart, only one ounce of cheese can provide as much as one-fourth of the officially recommended daily amount of calcium and one-fifth of our requirement of protein.

Cheese is fairly inexpensive, so there seems to be little excuse for any of us to be short on calcium or protein. Yet many children and young folks are just not getting enough. They substitute soft drinks for milk and do not add a bit of cheese to the meat they may eat for a lunch sandwich. Old folks develop osteoporosis or softening of the bones because they do not eat enough calcium-rich foods.

Riboflavin (vitamin B2) is another element which is very scarce in our daily diets. **Dairy products are good sources, but even so, one ounce of cheese gives you only up to 6 per cent of your daily requirement for this vitamin, which has so much to do with the health of your eyes, skin, digestive tract, mouth and gums.** The vitamin A in cheese comes from the cream, so it is not present in any quantity in the skimmed milk cheeses.

Another aspect of cheese which is sometimes worrisome for the health seeker is the salt content. For those of us who are trying to cut down on the amount of salt we use, some varieties of cheese are not recommended—not in any quantity at least. We do not have the sodium or salt content of some of the cheeses, but it is obvious to anyone eating Blue cheese, for example, or Roquefort, that there is a lot of salt in these varieties.

In other countries lots of cheese is made from the milk of other animals than cows. Goat milk makes fine cheese, as all health seekers know. Ewe's milk is made into cheese. Mare's milk and camel milk, yak milk and other kinds are popular in countries where these animals are plentiful.

"According to ancient records, cheese was used as a food source more than 4,000 years ago," reports *Cheese Varieties and Descriptions*, U.S. Department of Agriculture, Agriculture Handbook No. 54, which is available from the Superintendent of Documents, Washington, D. C. 20402. "It was made and eaten in Biblical times. Travelers from Asia are believed to have brought the art of cheesemaking to Europe. Cheese was made in many parts of the Roman Empire when it was at its height. Then cheesemaking was introduced to England by the Romans. During the Middle Ages as well as later, cheese was made and improved by the monks in the monasteries of Europe. . . . The Pilgrims included cheese when they made their famous voyage to America in the

Mayflower in 1620."

The USDA publication contains material on more than 400 cheeses, arranged alphabetically. Many of these are named after the town or community in which they are made. So, many cheeses with different local names are practically the same so far as taste and texture are concerned. And several altogether different kinds are known by the same local name. More than 800 names of cheeses are listed in the index of this 151-page book.

For those who want to know more about the nutritional value of different cheeses, an approximate analysis is given with every cheese where this was known. If you want to

Approximate Cholesterol and Salt Content of One Ounce of Cheese

Cheese	Choles-terol	Salt
Blue	21 mg.	–*
Camembert	20 mg.	–
Cheddar	28 mg.	170 mg.
Colby	26 mg.	–
Creamed cottage cheese	4 mg.	70 mg.
Uncreamed cottage cheese	2 mg.	–
Cream cheese	30 mg.	70 mg.
Edam	25 mg.	–
Mozzarella	15 mg.	–
Parmesan	20 mg.	170 mg.
Provolone	19 mg.	–
Swiss	24 mg.	170 mg.

(*We do not know the salt content of some of these)

compare the content of Cheddar cheese, cream cheese and Limburger cheese, for example, you would find that Cheddar is 25 per cent protein, 32 per cent fat, and 1.4 to 1.8 per cent salt. Cream cheese is 10 per cent protein, 35 to 38 per cent fat and 0.8 to 1.2 per cent salt. Limburger is 20 to 24 per cent protein, 26.5 to 29.5 per cent fat and 1.6 to 3.2 per cent salt.

If you are on a low-fat diet, cream cheese would be the least desirable for you. If you are trying to avoid salt, don't buy Limburger. In the case of cottage cheese, you are paying for a lot of moisture (up to 72 per cent), but much less fat for this cheese is made from skim milk. But it may be "creamed" before it is sold, so read the container to get the fat and protein content.

For a true cheese lover, this USDA booklet offers 151 pages of gustatory delight, for every cheese you ever heard of is here, with a description of how and where it is made, a bit of interesting history, and some secrets as to how its flavor and texture compare with that of other cheeses. For those who are writing about cheese, there's a fine bibliography of books and pamphlets on the subject.

Cottage cheese—mainstay of every reliable reducing diet and even some that are not so reliable, dainty addition to any fruit salad, high-protein fortification for the invalid's tray—is it really as good a food as we have been led to believe? You bet it is. And the housewife who does not include cottage cheese on her shopping list is missing out on one of the best and least expensive foods available.

In today's modern dairies, making tender, delicately flavored cottage cheese is an art. Skilled cheese makers produce various forms—from dry to creamed, from soft fine granules to large creamy curds. It is made from skimmed or defatted pasteurized milk, controlled amounts of lactic acid, with rennet and heat to coagulate the protein, called casein, into a soft curd. This curd is cut, the whey drained off, and, after a cold water wash, the cottage cheese is ready to be enjoyed. Last of all, sweet cream may be blended in for flavor. Cottage cheese, therefore, is a soft, uncured cheese made from skim milk, as we stated earlier.

Considering our calorie-counting attitudes, you should know that cottage cheese is made in different forms. If you are extremely weight conscious, listen to this: most cottage cheese sold is creamed, which means that the curd is mixed with cream to add flavor and smooth texture. This is creamed cottage cheese which usually contains 4 per cent butterfat. Salt is frequently added. If you count calories and watch salt, limit your intake of this type of cottage cheese. Uncreamed cottage cheese is a low-fat type which you may prefer.

Small-curd cottage cheese is known as old-fashioned: it has firm, small and tender curd making it very good for cooked dishes. Large curd or California style is easily mixed with other foods as a basis for salads. Pot style cheese is made from skim milk but may be salted or creamed, so read the label carefully, particularly the fat percentage where given. Bakers' cheese is the curd drained of whey but without cooking or washing. The curd may or may not be salted. This cheese is used in such bakery products as cheese cake, pie and pastries.

One-half cup of creamed cottage cheese gives about 120 calories and 17 grams of protein; one-half cup of uncreamed cheese yeilds about 108 calories and 22 grams of protein. Not too much difference, but those who want to lose pounds may prefer the less-fattening cheese.

One-half cup of creamed cottage cheese has almost as much protein as three ounces of cooked lean meat, fish or poultry. During warm days when you're just not up to a meat or poultry dish, you might make a cheese meal and keep up your valuable protein intake.

The curd carries most of the protein and much of the calcium and vitamin B2, reports *Food, Yearbook of Agriculture, 1959.* The solids carried off by the whey consist of the milk sugar (lactose), some of the protein, and a fairly large portion of vitamin B2 and the other B vitamins that were present in the milk originally.

Disproportionate losses of valuable nutrients in the whey

make it impossible to work out satisfactory nutritional equivalents between milk and cheese, except in terms of calcium or some other single nutrient, the USDA book explains. The amount of milk and cheese that would be equivalent in terms of any one nutrient would not necessarily be equivalent on the basis of other nutrients.

The flavor of cottage cheese deteriorates rapidly, *Food* says. It (and the other soft cheeses) should be stored—tightly covered—in the coldest part of the refrigerator.

Hard cheeses should be wrapped tightly before refrigerating to protect them from exposure to air. The original wrapping often is satisfactory, or any of the wide variety of wrapping materials available. Hard cheeses will keep indefinitely at refrigerator temperatures; exclusion of air is the important factor. Any mold that forms on the surface of hard cheese may be trimmed off before using.

Cheese spreads and cheese foods keep well without refrigeration until the container is opened. Refrigeration is advisable for the unused portion in an opened container. Of course, health seekers consider these cheeses atrocities.

Eleven ounces or about one and one-half cups of cottage cheese will supply about the same amount of calcium as an 8-ounce glass of whole milk, according to *Food*. Also, one-inch cube of Cheddar-type cheese equals two-thirds cup of milk; 2 tablespoons of cream cheese equal 1 tablespoon of milk.

As we mentioned, any surface mold that develops on hard natural cheese should be trimmed off before eating. In Blue cheese or Roquefort, mold is an important part of the cheese flavor and should, of course, be eaten. Hard cheese that has dried out can be grated and stored in a tightly covered jar for use in casseroles, omelets, etc.

Don't freeze cheese unless you don't mind the crumbly texture which develops. This is, of course, acceptable in Blue or Roquefort cheese. And some hard varieties can be frozen with satisfactory results, in small pieces, one pound or less, not more than one inch thick: Brick, Cheddar, Edam, Gouda,

Muenster, Port du Salut, Swiss, Provolone, Mozzarella and Camembert. Wrap tightly in freezer paper or foil, freeze quickly at zero degrees and store no longer than six months. Thaw slowly, preferably in the refrigerator.

Cheese and fruit make the best possible dessert, for cheese is rich in a number of nourishing factors which are not so abundant in fruit: Calcium, protein and the B vitamins, for example.

Versatile cottage cheese is good atop pancakes with jelly, at lunch in sandwiches, as dessert for dinner. Use cottage cheese with spices and seasonings, with fruit, vegetables, meat, fish, eggs.

Did you ever try stuffing tomatoes with cottage cheese and baby green limas? Or, as a topping treat—chopped pistachio nuts sprinkled generously over cottage cheese? On a busy day, try serving hot potatoes, boiled in their jackets, split, seasoned, and topped with creamy cottage cheese at room temperature. Here's a healthy favorite your family will love: whip a carton of creamed cottage cheese (or any required amount) in a blender or electric mixer until the cheese is smooth and light. Spoon onto gingerbread or warm spice cake squares and dust with cinnamon. Looks pretty, tastes wonderful!

For your next luncheon, go Hawaiian. Serve pineapple halves filled with creamy, luscious chive cottage cheese and fresh fruit.

Chive cottage cheese takes this salad out of the ordinary class and into the exotic. The fruit tastes richer and more deliciously poignant when teamed with this elite of cottage cheese. The chives give additional delicate flavor which brings new interest to a fruit salad, making it a very special dish.

Although chive cottage cheese is a wonderful boon to calorie watchers, because of its distinctive flavor, it is also very popular with those who don't need to watch their weight.

Don't for a moment think that a slice of cheese on a piece

of bread or a cracker is the only way you can eat cheese. It is one of the most versatile of foods and can be used in many recipes where it enriches the entire dish with its protein, calcium and B vitamins.

Your health food store and/or your book store will have many recipe books using cheese. A very useful 28-page booklet, *Cheese in Family's Meals*, is available for 20 cents from the Superintendent of Documents, Washington, D. C. 20402. Ask for Home and Garden Bulletin, No. 112, U. S. Department of Agriculture.

Index

D

E

F

G

H

I

Read What the Experts Say About Larchmont Books!

The Complete Home Guide to All the Vitamins

"This is a handy book to have at home, for it discusses in clear, simple language just what vitamins are, why we need them, and how they function in the body."—*Sweet 'n Low*

"Want to know what vitamins you need and why? Then this is your cup of tea. A paperback that tells you everything you ever wanted to know about vitamins and maybe were afraid to ask... Read it and reap."—*Herald American and Call Enterprise, Allentown, Pa.*

Minerals: Kill or Cure?

"Written both for professional and non-professional readers, this book offers excellent background for additional discoveries that are inevitable in the next few years..."—*The Total You*

Eating in Eden

"This book contains very valuable information regarding the beneficial effects of eating unrefined foods..."—*Benjamin P. Sandler, M.D., Asheville, N.C.*

"We must be reminded again and again what junk (food) does and how much better we would be if we avoided it. This book serves to do this."—*A. Hoffer, M.D., Ph.D.*

Read What the Experts Say About Larchmont Books!

Megavitamin Therapy

"This book provides a much-needed perspective about the relationship of an important group of medical and psychiatric conditions, all of which seem to have a common causation (the grossly improper American Diet) and the nutritional techniques which have proven to be of great benefit in their management."
—Robert Atkins, M.D., author of "Dr. Atkins' Diet Revolution," New York.

"This responsible book gathers together an enormous amount of clinical and scientific data and presents it in a clear and documented way which is understandable to the average reader . . . The authors have provided critical information plus references for the acquisition of even more essential knowledge."*—David R. Hawkins, M.D., The North Nassau Mental Health Center, Manhasset, New York.*

Health Foods

"This book (and "Is Low Blood Sugar Making You a Nutritional Cripple") are companion books worth adding to your library. The fact that one of the books is labeled "health foods" is an indication how far our national diet has drifted away from those ordinary foods to which man has adapted over the past million years. . . ."*—A. Hoffer, M.D., Ph.D., The Huxley Newsletter.*

"A sensible, most enlightening review of foods and their special qualities for maintenance of health . . . ."*—The Homeostasis Quarterly.*

Read What the Experts Say About Larchmont Books!

Body, Mind and the B Vitamins

"I feel that "Body, Mind and the B Vitamins" is an excellent, informative book. I recommend everyone buy two copies; one for home and one to give to their physician."—*Harvey M. Ross, M.D., Los Angeles, Calif.*

Program Your Heart for Health

"What is unique about this book is that the tremendous body of fascinating information has been neatly distilled so that the problems and the solutions are quite clear. . . . (This book) will be around for a long time . . . so long as health continues to be the fastest growing failing business in the United States and so long as it is not recognized that the medical problem is not medical but social."—*E. Cheraskin, M.D., D.M.D., Birmingham, Ala.*

"If more people were to read books such as this one and were to institute preventive medical programs early in life, the mortality in heart disease would drop precipitously as well as in our other serious medical problems."—*Irwin Stone, Ph.D., San Jose, Calif.*

"**Program Your Heart for Health**" contains a wealth of data. I plan to make use of it many times."—*J. Rinse, Ph.D., East Dorset, Vt.*

Larchmont Preventive Health Library

The Library will consist of the following books, issued as indicated. For quick reference, we have left off the full title of each book, which is "Improving Your Health with Vitamin A," etc.

1978

1. Vitamin A
2. Niacin (Vitamin B3)
3. Vitamin C
4. Vitamin E
5. Calcium and Phosphorus
6. Zinc

1979

7. Thiamine (B1) and Riboflavin (B2)
8. Pyridoxine (B6)
9. Iodine, Iron and Magnesium
10. Sodium and Potassium
11. Copper, Chromium and Selenium
12. Vitamin B12 and Folic Acid

1980

13. Vitamin D and Vitamin K
14. Pantothenic Acid
15. Biotin, Choline, Inositol and PABA
16. Protein and Amino Acids
17. Natural Foods
18. Other Trace Minerals